FUNDING SMITHSONIAN SCIENTIFIC RESEARCH

Committee on Smithsonian Scientific Research

Board on Life Sciences

Division on Earth and Life Studies

Board on Physics and Astronomy

Division on Engineering and Physical Sciences

NATIONAL RESEARCH COUNCIL
OF THE NATIONAL ACADEMIES

THE NATIONAL ACADEMIES PRESS
Washington, D.C.
www.nap.edu

THE NATIONAL ACADEMIES PRESS • 500 FIFTH STREET, N.W. • Washington, DC 20001

NOTICE: The project that is the subject of this report was approved by the Governing Board of the National Research Council, whose members are drawn from the councils of the National Academy of Sciences, the National Academy of Engineering, and the Institute of Medicine. The members of the committee responsible for the report were chosen for their special competences and with regard for appropriate balance.

This study was supported by Grant No. T0236CC10348 between the National Academies and the Smithsonian Institution. Any opinions, findings, conclusions, or recommendations expressed in this publication are those of the author(s) and do not necessarily reflect the views of the organizations or agencies that provided support for the project.

International Standard Book Number 0-309-08633-7

Additional copies of this report are available from the National Academies Press, 500 Fifth Street, N.W., Lockbox 285, Washington, DC 20055; (800) 624-6242 or (202) 334-3313 (in the Washington metropolitan area); Internet, http://www.nap.edu

COVER: Clockwise from top left corner. Photos reprinted with permission from the Smithsonian Institution.

Mei Xiang, the female giant panda at the National Zoological Park. Mei Xiang, along with the male Panda Tian Tian, are on a 10-year loan, under an agreement with the China Wildlife Conservation Association. Photo by Jessie Cohen.

X-ray image of the Crab Nebula, the remnant of a supernova explosion in 1054 A.D., taken by the Chandra X-ray Observatory, which is operated by the Smithsonian Astrophysical Observatory's Chandra X-ray Center. Photo credit: NASA/CXC/SAO.

Spacesuit preservation by staff at the Smithsonian Center for Materials Research and Education. Photo by Doc Dougherty.

View of the Smithsonian Tropical Research Institute in Barro Colorado Island, Panama. Photo by Marcos Guerra.

Feather identification expert Roxie Laybourne, amidst a portion of bird collection at the National Museum of Natural History. Photo by Chip Clark.

View of the Smithsonian Environmental Research Center in Edgewater, Maryland. Photo by Mark Haddon.

Printed in the United States of America

THE NATIONAL ACADEMIES

Advisers to the Nation on Science, Engineering, and Medicine

The **National Academy of Sciences** is a private, nonprofit, self–perpetuating society of distinguished scholars engaged in scientific and engineering research, dedicated to the furtherance of science and technology and to their use for the general welfare. Upon the authority of the charter granted to it by the Congress in 1863, the Academy has a mandate that requires it to advise the federal government on scientific and technical matters. Dr. Bruce Alberts is president of the National Academy of Sciences.

The **National Academy of Engineering** was established in 1964, under the charter of the National Academy of Sciences, as a parallel organization of outstanding engineers. It is autonomous in its administration and in the selection of its members, sharing with the National Academy of Sciences the responsibility for advising the federal government. The National Academy of Engineering also sponsors engineering programs aimed at meeting national needs, encourages education and research, and recognizes the superior achievements of engineers. Dr. Wm. Wulf is president of the National Academy of Engineering.

The **Institute of Medicine** was established in 1970 by the National Academy of Sciences to secure the services of eminent members of appropriate professions in the examination of policy matters pertaining to the health of the public. The Institute acts under the responsibility given to the National Academy of Sciences by its congressional charter to be an adviser to the federal government and, upon its own initiative, to identify issues of medical care, research, and education. Dr. Harvey V. Fineberg is president of the Institute of Medicine.

The **National Research Council** was organized by the National Academy of Sciences in 1916 to associate the broad community of science and technology with the Academy's purposes of furthering knowledge and advising the federal government. Functioning in accordance with general policies determined by the Academy, the Council has become the principal operating agency of both the National Academy of Sciences and the National Academy of Engineering in providing services to the government, the public, and the scientific and engineering communities. The Council is administered jointly by both Academies and the Institute of Medicine. Dr. Bruce Alberts and Dr. Wm. Wulf are chairman and vice chairman, respectively, of the National Research Council.

www.national-academies.org

COMMITTEE ON SMITHSONIAN SCIENTIFIC RESEARCH

CORNELIUS J. PINGS (Chair), Association of American Universities (Emeritus), Pasadena, CA
BARBARA L. BEDFORD, Department of Natural Resources, Cornell University, Ithaca, NY
MARC DAVIS, Department of Astronomy, University of California, Berkeley, CA
HUGH W. DUCKLOW, School of Marine Science, The College of William and Mary, Gloucester Point, VA
JONATHAN FINK, Office of Vice President for Research and Economic Affairs, Arizona State University, Tempe, AZ
ANTHONY C. JANETOS, H. John Heinz III Center for Science, Economics and the Environment, Washington, DC
KENNETH I. KELLERMANN, National Radio Astronomy Observatory, Charlottesville, VA
J. PATRICK KOCIOLEK, Diatom Collection, California Academy of Sciences, San Francisco, CA
DANIEL A. LIVINGSTONE, Department of Biology, Duke University, Durham, NC
MICHAEL J. NOVACEK, Office of Senior Vice-President and Provost, and Division of Paleontology, American Museum of Natural History, New York, NY
BRUCE A. RIDEOUT, Department of Pathology, Zoological Society of San Diego, San Diego, CA
ETHAN SCHREIER, Associated Universities, Inc., Washington, DC
PATRICIA WRIGHT, Department of Anthropology, State University of New York, Stony Brook, NY

Staff

Frances E. Sharples, Director, Board on Life Sciences
Donald C. Shapero, Director, Board on Physics and Astronomy
Evonne Tang, Study Co-director
Michael Moloney, Study Co-director
Valerie Gutmann, Project Assistant
Nelson Quinones, Project Assistant
Norman Grossblatt, Senior Editor

BOARD ON LIFE SCIENCES

BOARD ON PHYSICS AND ASTRONOMY

JOHN P. HUCHRA (Chair), Harvard-Smithsonian Center for Astrophysics, Cambridge, MA
ROBERT C. RICHARDSON (Vice Chair), Cornell University, Ithaca, NY
JONATHAN BAGGER, The Johns Hopkins University, Baltimore, MD
GORDON A. BAYM, University of Illinois at Urbana-Champaign, Urbana, IL
CLAUDE R. CANIZARES, Massachusetts Institute of Technology, Cambridge, MA
WILLIAM EATON, National Institutes of Health, Bethesda, MD
WENDY L. FREEDMAN, Carnegie Observatories, Pasadena, CA
FRANCES HELLMAN, University of California at San Diego, La Jolla, CA
KATHRYN LEVIN, University of Chicago, James Franck Institute, Chicago, IL
CHUAN LIU, University of Maryland, College Park, MD
LINDA J. MAGID, University of Tennessee, Knoxville, TN
THOMAS M. O'NEIL, University of California at San Diego, La Jolla, CA
JULIA M. PHILLIPS, Sandia National Laboratories, Albuquerque, NM
BURTON RICHTER, Stanford University, Menlo Park, CA
ANNEILA I. SARGENT, California Institute of Technology, Pasadena, CA
JOSEPH H. TAYLOR, JR., Princeton University, Princeton, NJ
THOMAS N. THEIS, IBM, Yorktown Heights, NY
CARL E. WIEMAN, JILA/University of Colorado, Boulder, CO

Staff

Donald C. Shapero, Director

Preface

The superb system of federal support for US science and technology has produced 5 decades of discovery and innovation that have changed the way we live. This use of public resources is widely agreed to be a great public good. Nevertheless, it is the legitimate responsibility of government managers to try to ensure that public investments in science and technology are used wisely and continue to produce high-quality results.

Fulfillment of that responsibility underlies the preparation of this report, which grew out of a concern that scientific research conducted by parts of the Smithsonian Institution might not be the best obtainable because a portion of the funding for this research is directly appropriated by the federal government without peer-reviewed competition.

The Smithsonian Institution is one of the nation's most venerable organizations. Founded in the mid-19th century with a substantial bequest to the US government from the English scientist James Smithson, almost 150 years later it still carries out the mission Smithson assigned it—"the increase and diffusion of knowledge." But even venerable institutions should be well managed and the quality of their work ensured. At the request of the Office of Management and Budget, therefore, the Smithsonian asked the National Academy of Sciences, in partnership with the National Academy of Public Administration (NAPA), to evaluate whether the federal research funding now given by direct appropriation to the Smithsonian's scientific programs could be better invested by transferring these funds to the National Science Foundation (NSF) to further support its competitively awarded research grants programs.

In response, the National Research Council of the National Academies established the Committee on Smithsonian Scientific Research. The persons appointed to serve on the Committee have a wealth of experience in the scientific fields represented in the Smithsonian's own scientific portfolio and have substantial knowledge of the Smithsonian science units themselves and their research output. Expertise represented on the Committee includes astronomy and astrophysics, ecology, tropical and marine biology, biodiversity conservation, veterinary medicine, anthropology, paleontology, biogeochemistry, volcanology, systematics, and the collection and preservation of museum specimens. To carry out its task, the full Committee met twice, at the beginning and end of the project. In between, its members, divided into three panels, met frequently by teleconference to draft this report and provide the basis of its recommendations. The report represents the consensus of the Committee's views. Throughout its preparation, we remained in close touch with our NAPA counterparts to ensure that our two reports were well coordinated.

It should be noted that the terms of the possible transfer of funds from the Smithsonian to the NSF were not specified to the Committee. The Committee, therefore, was forced to design reasonable scenarios for how such a transfer might be made. The Committee elected to evaluate an array of cases, including those in which all funding, including salary support for Smithsonian scientists, would be transferred and cases in which NSF would be directed to use the transferred funds to support research in the same disciplinary fields as before or even to maintain the programs and operations of the Smithsonian facilities more or less intact. The impacts of a funding transfer would vary considerably according to the terms established and in some cases could be draconian for the well-being of the Smithsonian scientific staff and programs.

I wish to thank all the members of the Committee for their valuable contributions and for their insights into the scientific and societal issues surrounding this project. I also wish to acknowledge the National Research Council staff (Evonne Tang, Michael Moloney, Fran Sharples, and Don Shapero) for their thorough and thoughtful assistance with all aspects of the preparation of this report.

Cornelius J. Pings, *Chair*
Committee on Smithsonian Scientific Research

Acknowledgments

This report has been reviewed in draft form by persons chosen for their diverse perspectives and technical expertise, in accordance with procedures approved by the National Research Council's Report Review Committee. The purpose of this independent review is to provide candid and critical comments that will assist the institution in making its published report as sound as possible and to ensure that the report meets institutional standards for objectivity, evidence, and responsiveness to the study charge. The review comments and draft manuscript remain confidential to protect the integrity of the deliberative process. We wish to thank the following individuals for their review of this report.

John D. Aber, University of New Hampshire
Charlie Costain, Canadian Institute of Conservation
W. Gary Ernst, Stanford University
Christopher B. Field, Carnegie Institution of Washington
Wendy L. Freedman, Observatories of the Carnegie Institution
William B. Karesh, Wildlife Conservation Society
Patrick V. Kirch, University of California, Berkeley
Neal F. Lane, Rice University
Stuart Pimm, Duke University
Burton Richter, Stanford University
Anneila I. Sargent, California Institute of Technology
Andrew R. Solow, Woods Hole Oceanographic Institution
Ivan Valiela, Boston University
Terry L. Yates, University of New Mexico

Although the reviewers listed above have provided many constructive comments and suggestions, they were not asked to endorse the conclusions or recommendations, nor did they see the final draft of the report before its release. The review of this report was overseen by Michael T. Clegg, University of California, Riverside, and Louis J. Lanzerotti, Bell Laboratories, Lucent Technologies. Appointed by the National Research Council, they were responsible for making certain that an independent examination of this report was carried out in accordance with institutional procedures and that all review comments were carefully considered. Responsibility for the final content of this report rests entirely with the authoring committee and the institution.

Contents

APPENDIXES

Executive Summary

The Smithsonian Institution (SI) was established as an independent trust instrumentality in 1846 dedicated to "the increase and diffusion of knowledge among men" as laid out in James Smithson's bequest to the US government. To accomplish its mission, the Smithsonian throughout its history has combined high quality research conducted by its scientific research centers with public outreach through exhibitions of its collections in museums. Although the Smithsonian's science centers and their research are highly regarded by the scientific community, they are much less well known to the general public than their museums.

The Smithsonian Institution receives an annual federal appropriation toward its operating costs, which includes funds in support of research at the Smithsonian. In the FY 2003 presidential budget, the Office of Management and Budget (OMB) called for a review "to recommend how much of the funds directly appropriated to the Smithsonian for scientific research should be awarded competitively," and proposed to transfer these funds to the National Science Foundation (NSF). Specifically, OMB expressed concern about the Smithsonian's classification of its allocation of federal research funds as "inherently unique"—that is, research programs that are funded without competition.

The apparent absence of competition in the Smithsonian science centers raises concerns about a lack of quality assurance in Smithsonian research. Moreover, it is fair to ask whether the federal support given to the Smithsonian's science programs could be used more effectively for science if the funds were awarded through a competitive process open to all researchers. After the release of the budget document, the Smithsonian commissioned reviews by the National Academy of Sciences (NAS) and the National Academy of Public Administration (NAPA) to address the questions raised by the OMB. This is the report of the NAS review; the NAPA study will be the subject of a separate report.

The Committee on Smithsonian Scientific Research was charged to provide specific recommendations and a rationale with criteria on what parts of the Smithsonian's research portfolio should continue to be exempt from priority setting through competitive peer-reviewed grant programs because of *uniqueness* or *special contributions*. The charge to the Committee called for a review of the scientific research centers that report to the Smithsonian's Under Secretary for Science—the National Museum of Natural History, the Smithsonian Astrophysical Observatory, the National Zoological Park, the Smithsonian Tropical Research Institute, the Smithsonian Center for Materials Research and Education, and the Smithsonian Environmental Research Center. The Committee was also charged to consider the effects on the Smithsonian, the research centers, and the relevant scientific fields of re-allocating the current federal support to a competitive process. Finally, the Committee was asked to make recommendations on how any Smithsonian science programs that continued to receive direct federal appropriations should be regularly evaluated and compared with other research in the relevant fields. The Committee was not asked to review the funding of SI research centers that report to the Smithsonian's Under Secretary for American Museums and National Programs.

To respond to its charge, the Committee examined the research programs and the funding structure at the six Smithsonian scientific research centers. It also considered possible consequences of removing direct federal appropriations to the Smithsonian science programs and reallocating the funds to open competition.

In carrying out its review, the Committee established a framework of criteria to be applied to its review of the Smithsonian research centers in the execution of its task. The Committee considered

- The nature of the Smithsonian as a scientific institution.
- How *uniqueness* and *special contribution* apply to each of the six science centers covered by the study. In the context of this study, *uniqueness* and *special contribution* may have many meanings that refer to special attributes associated with a particular research center.
- How opening some of or all the support now given to each of the centers to a competitive process would affect the science involved.
- How the centers might be evaluated regularly to ensure that the quality of their science is maintained if any of the six are deemed to be unique and to warrant continuation of the current system of support.

The six research centers, taken together, embody SI's research program and constitute the mechanism whereby SI carries out its charter to increase and diffuse knowledge. The Committee considered the work of each SI unit, its role and status in the scientific enterprise, and whether the terms *uniqueness* and *special contribution* should be applied to its research. In arriving at its findings, conclusions, and recommendations, the Committee drew on information received from, and interviews with, representatives of the central offices of the Smithsonian and the research centers, on the expertise and relevant knowledge of the Committee members themselves, and on informal contact with members of the wider scientific community.

FINDINGS AND CONCLUSIONS

A: The research performed by the National Museum of Natural History, the National Zoological Park, and the Smithsonian Center for Materials Research and Education is inextricable from their missions and is appropriately characterized by the terms *unique* and *special contributions*.

B: The Smithsonian Astrophysical Observatory, the Smithsonian Environmental Research Center, and the Smithsonian Tropical Research Institute are world-class scientific institutions that combine facilities, personnel, and opportunities for specialized long-term research that is enabled by the stability of federal support. These units are engaged in research that supports the mission of the Smithsonian Institution as a whole—increasing knowledge and providing supporting expertise for the activities of other SI units, including educational activities.

C: Funding for research at the Smithsonian's research centers comes from a mix of sources, including a substantial fraction received through open competitive programs.

D: The Smithsonian Institution plays an important role in the overall US research enterprise and contributes to the healthy diversity of the nation's scientific enterprise.

E: Mechanisms at the Smithsonian scientific research centers for evaluating overall scientific productivity and for evaluating the productivity of individual scientists are variable and inconsistent.

F: Communication between the research centers and the central management of the Smithsonian Institution appears to be weak.

Consequences of Transferring Federally Appropriated Research Funds from the Smithsonian

The following findings and conclusions stem from the Committee's consideration of the consequences of reallocating the federal funds appropriated currently to the Smithsonian to a competitively peer-reviewed program at NSF.

G: In general, transfer of all federal research funds (including salary and, in some cases, infrastructure support) would greatly reduce and possibly eliminate the role of the federal government in the long-term support of the core scientific research staff who provide the foundation of the Smithsonian research program. A withdrawal of federal support of this magnitude would make maintaining the staff and programs of the centers extremely difficult and would very likely lead to the demise of much of the Smithsonian's scientific research program.

H: Transferring the federally appropriated research funds for the National Museum of Natural History and the National Zoological Park to competitive programs at the National Science Foundation is likely to jeopardize their standing in the museum and zoo communities and could seriously damage aspects of their nonresearch roles. If the fund transfer were large and included salary support, the positions of critical museum and zoo personnel could be threatened. Loss of core funds could also lead to the closure of the Smithsonian Center for Materials Research and Education.

I: Transferring directly appropriated funds from the Smithsonian Astrophysical Observatory, the Smithsonian Environmental Research Center, and the Smithsonian Tropical Research Institute to a competitive mechanism while trying to maintain the centers in the Smithsonian could produce consequences ranging from moderately or seriously deleterious to termination of their operations.

J: The Committee could not identify any substantial advantages with respect to organization, management, or quality assurance that would accrue from changing the current system of federally appropriated research funding for the Smithsonian Astrophysical Observatory, the Smithsonian Environmental Research Center, and the Smithsonian Tropical Research Institute.

K: The Committee identified little or no scientific benefit of transferring federal funds away from the Smithsonian. The implications for the relevant scientific fields are likely to be adverse.

L: The broad mission of the Smithsonian Institution would be compromised if the links between the Smithsonian and its research centers were broken by transferring sponsorship of the centers to the National Science Foundation.

RECOMMENDATIONS

1. Research is an intrinsic part of the mission of the National Museum of Natural History and the National Zoological Park. These centers should continue to be exempt from open competition for research funding because of the uniqueness and special contributions conferred by association with their collections.

2. The Smithsonian Center for Materials Research and Education occupies a highly specialized research niche that is of unique and major value to museums of the Smithsonian Institution and to the museum community at large. Hence, the Committee believes that the center should continue to be exempt from open competition for research funding because of its uniqueness and special contributions to the museum community.

3. The Committee believes that the Smithsonian Astrophysical Observatory, the Smithsonian Tropical Research Institute, and the Smithsonian Environmental Research Center should continue to receive federally appropriated research funding. Use of public funds by these facilities is already producing science of the highest quality. Much of the "research funding" (for other than salary and infrastructure costs) is already obtained via competition. Any benefits of shifting these three facilities to the jurisdiction of another organization would be greatly outweighed by the harm done to their contributions to the relevant scientific fields.

4. Regular in-depth reviews by external advisory committees are essential for maintaining the health, vitality, and scientific excellence of the Smithsonian Institution. Although details of the nature and processes of the reviews may vary to accommodate differences among the six centers, such institutional reviews should be uniformly required for all six Smithsonian science centers and for their individual departments, if warranted by their size. Retrospective external peer review is especially important for areas not routinely engaging in competition for grants and contracts. Regular cycles of review followed by strategic planning offer the best means of ensuring that the quality of SI's science is maintained.

5. The research programs at the Smithsonian Institution provide essential support to the museums and collections, make substantial contributions to the relevant scientific fields, and fulfill the broader Smithsonian mission to "increase and diffuse knowledge." The Committee urges a stronger sense of institutional stewardship for these research programs as integral components of the Smithsonian. The Secretary and the Board of Regents should improve communication with the research centers and become strong advocates for their goals and achievements in a manner that is compelling to the Executive Branch, Congress, and the public.

1

Introduction and Background

ORIGINS OF THE SMITHSONIAN INSTITUTION[1]

The Smithsonian Institution (SI) originated in the mind of the English scientist James Smithson. Before his death in 1829, he named the United States the trustee of a sizable sum of money on the condition that the United States establish a research and educational institution to benefit all people. Congress accepted the trust in 1836 and debated what type of institution the Smithsonian should be for the next 10 years. In 1846, Congress and President James Polk approved a statute establishing the Smithsonian as an institution for "the increase and diffusion of knowledge among men," as envisioned in Smithson's will. SI is unlike any other federal organization in that it is an independent trust instrumentality, a product of the United States government that has no governing function.

Today, SI comprises 16 museums and gallery buildings, the National Zoological Park, and several research centers. (Figure 1-1 shows the SI organization chart.) Throughout its history, the balance of the Smithsonian's focus between scientific research and natural history and museum collections has changed under the influence of the various men who have served as Secretary and their visions for the institution. At the time of its founding in 1846, the Board of Regents, the institution's governing body, sought as Secretary a person with "eminent scientific and general requirements" who might take on the task of "advancing science

[1] Information on the history of the Smithsonian Institution and its research facilities was obtained from the Institutional History Division of the Smithsonian Archives.

and promoting letters by original research and effort." The Secretary was also expected "to act as a respected channel of communication between the institution and scientific and literary individuals and societies in this and foreign countries." The regents chose Joseph Henry, who might have been America's most distinguished scientist at the time. Henry later served as second president of the National Academy of Sciences, which he had helped President Lincoln to establish.

Henry strongly promoted research as the key focus of SI. Although the act of Congress establishing SI directed the Institution to have a library, a museum, and an art gallery, Henry believed that it should have such a charge only temporarily and that the management and operation of these entities should be transferred to other agencies as soon as possible. However, through the efforts of the then Assistant Secretary Spencer Fullerton Baird, the Smithsonian began to receive major natural history and cultural collections that document the minerals, fossils, rocks, animals, and plants of North America, which ultimately grew to be the best collection of its kind in the world. Baird succeeded Henry as Secretary in 1878 and embraced the museum mandate he favored for the Institution. Under Baird, research and public activities centered around natural history.

In 1887, Samuel Pierpont Langley, a prominent and internationally respected scientist in astrophysics and aeronautics, was appointed the third Secretary of SI. Under his leadership, the balance of Smithsonian interests tilted toward the physical sciences. Langley established the Smithsonian Astrophysical Observatory and the departments of biology, anthropology, and geology.

In 1907, Charles Doolittle Walcott, one of the leading paleontologists of the time, began a 20-year term as Secretary. The return to prominence of natural history research at SI culminated in the opening of the National Museum of Natural History to the public in 1910. Under Walcott's leadership, the Smithsonian also participated in a major biological survey in the Panama Canal Zone, an effort that led to the establishment of the Smithsonian Tropical Research Institute.

For the rest of the 20th century, SI maintained its high standing in the advancement of science under the direction of scientist-secretaries. SI science expanded rapidly in the 1960s and 1970s, when ample funds were available for equipment, expeditions, and collection management. Those favorable circumstances attracted world-class scientists to the Institution. In 1965, Secretary S. Dillon Ripley established the newest of the federally supported science units, the Chesapeake Bay Center for Environmental Studies, now known as the Smithsonian Environmental Research Center, for the conduct of natural history and ecological research. As universities became less interested in whole-organism study, the Smithsonian, with

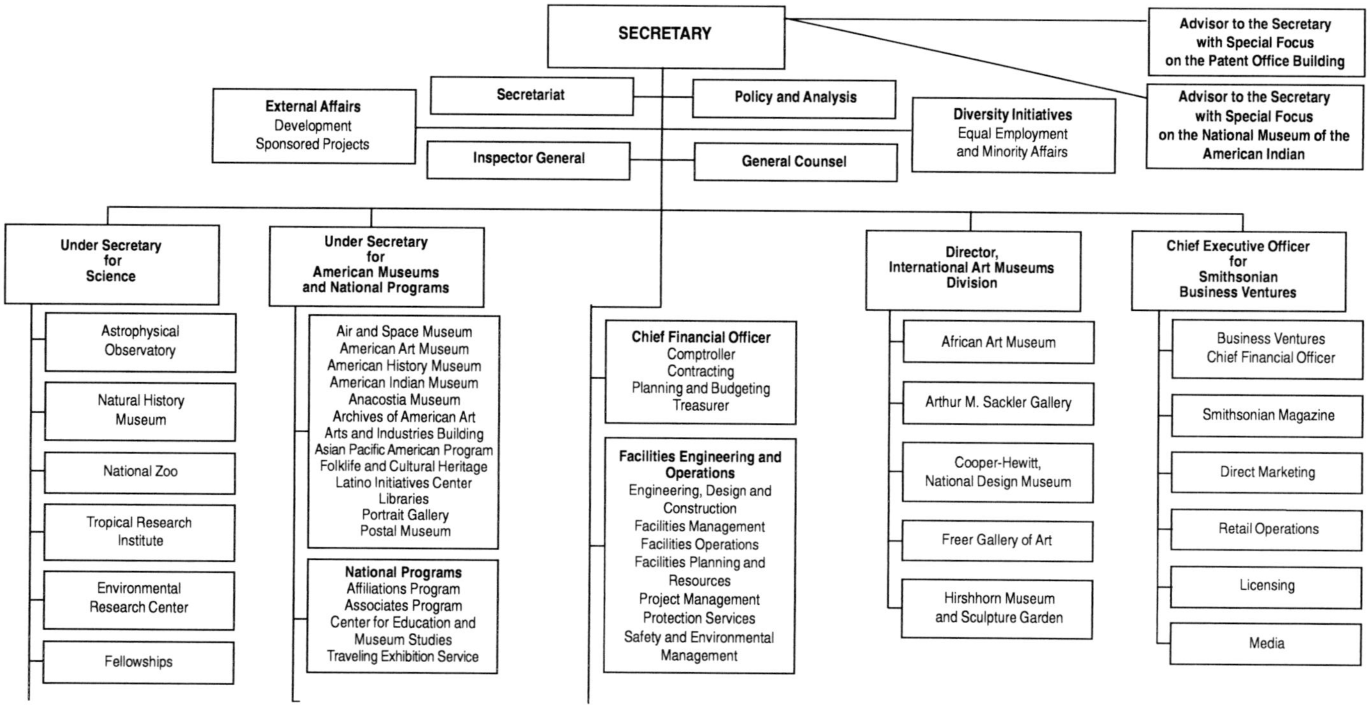
SECRETARY
Advisor to the Secretary with Special Focus on the Patent Office Building
Advisor to the Secretary with Special Focus on the National Museum of the American Indian
External Affairs
Development
Sponsored Projects
Secretariat
Policy and Analysis
Inspector General
General Counsel
Diversity Initiatives
Equal Employment and Minority Affairs
Under Secretary for Science
Astrophysical Observatory
Natural History Museum
National Zoo
Tropical Research Institute
Environmental Research Center
Fellowships
Under Secretary for American Museums and National Programs
Air and Space Museum
American Art Museum
American History Museum
American Indian Museum
Anacostia Museum
Archives of American Art
Arts and Industries Building
Asian Pacific American Program
Folklife and Cultural Heritage
Latino Initiatives Center
Libraries
Portrait Gallery
Postal Museum
National Programs
Affiliations Program
Associates Program
Center for Education and Museum Studies
Traveling Exhibition Service
Chief Financial Officer
Comptroller
Contracting
Planning and Budgeting
Treasurer
Facilities Engineering and Operations
Engineering, Design and Construction
Facilities Management
Facilities Operations
Facilities Planning and Resources
Project Management
Protection Services
Safety and Environmental Management
Director, International Art Museums Division
African Art Museum
Arthur M. Sackler Gallery
Cooper-Hewitt, National Design Museum
Freer Gallery of Art
Hirshhorn Museum and Sculpture Garden
Chief Executive Officer for Smithsonian Business Ventures
Business Ventures Chief Financial Officer
Smithsonian Magazine
Direct Marketing
Retail Operations
Licensing
Media

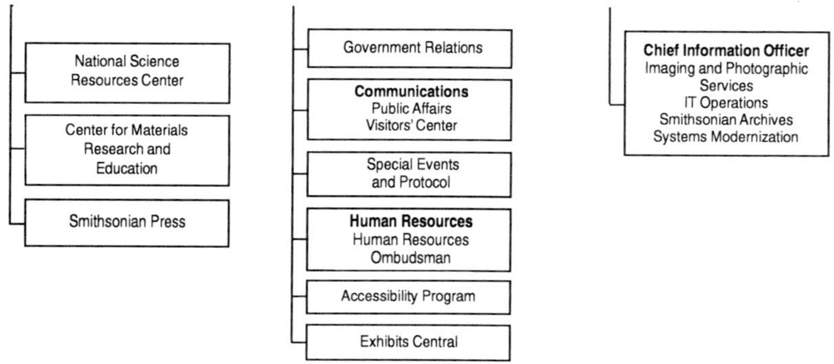

Note: Many office names have been abbreviated to conserve space. Please check with an individual unit to confirm its official name.

FIGURE 1-1 Organization chart of the Smithsonian Institution.

its collections and its research centers on protected land, was able to take the lead in research that focused on long-term, large-scale data-gathering in terrestrial and marine ecology, global change, and biodiversity.

DEVELOPMENT OF THE SCIENTIFIC RESEARCH CENTERS

Although there are many science-related museums renowned for their roles in public education around the world, their associated research centers are often less visible to the public. This is true of the Smithsonian: the general public is mostly unaware of the scientific and other research conducted by the Institution. Because of the lack of understanding of the continuing and central role of research in the mission of the Smithsonian and the misconception that the Institution is solely a collection of museums and a zoo, justifying a substantial budget for research at the Institution to policy-makers can be difficult. Indeed, in his formal inaugural address as the 11th Secretary of SI, Lawrence Small decried the lack of awareness of SI science among the public, members of Congress, and the administration.

Today, six SI units are assigned to the management of the Under Secretary for Science as "scientific research centers." These centers include the most publicly familiar entities—the National Museum of Natural History (NMNH) in Washington, DC, and the National Zoological Park (NZP) in Washington, DC, and Front Royal, Virginia. The others are the Smithsonian Astrophysical Observatory (SAO) in Cambridge, Massachusetts; the Smithsonian Tropical Research Institute (STRI) in Panama; the Smithsonian Center for Materials Research and Education (SCMRE) in Suitland, Maryland; and the Smithsonian Environmental Research Center (SERC) in Edgewater, Maryland. SI also operates several other research centers, such as the Center for Earth and Planetary Sciences in the National Air and Space Museum and various research units in its art museums; these research centers report to the Under Secretary for American Museums and National Programs.

The act of Congress that established the Smithsonian in 1846 (9 Stat 102) specifically provided for a natural history museum. It stated that a building should be constructed "with suitable rooms or halls for the reception and arrangement, upon a liberal scale, of objects of natural history, including a geological and mineralogical cabinet" and that

> as suitable arrangements can be made for their reception, all objects of art and of foreign and curious research, and all objects of natural history, plants, and geological and mineralogical specimens, belonging, or hereafter to belong, to the United States, which may be in the city of Washington, in whosesoever custody the same may be, shall be delivered to such persons as may be authorized by the board of regents to receive

> them, and shall be arranged in such order, and so classed, as best [to] facilitate the examination and study of them, in the building so as aforesaid to be erected for the institution; and the regents of said institution shall afterwards, as new specimens in natural history, geology, or mineralogy, may be obtained for the museum of the institution, by exchanges of duplicate specimens belonging to the institution, (which they are hereby authorized to make,) or by donation, which they may receive, or otherwise, cause such new specimens to be also appropriately classed and arranged.

The relationship between the collections acquired by other government offices and what was called the National Museum at the Smithsonian was reinforced in 1879 legislation that created the US Geological Survey (USGS; 20 Stat 377). It required that all natural history collections made by the US government "when no longer needed for investigations in progress shall be deposited in the National Museum." A separate building was erected for the natural history collections and opened to the public in 1910; it was renamed the National Museum of Natural History in 1969. Box 1-1 presents a concise chronology of the development of the Smithsonian.

SI Secretary Samuel P. Langley established SAO in 1890. An astrophysicist himself, Langley set up SAO in Washington, DC, primarily for studies of the sun, using Smithsonian trust funds. A year after its establishment, Congress made its first appropriation, totaling $10,000 for FY 1892, dedicated to the maintenance of the observatory. SAO is now in Cambridge, Massachusetts, where it moved from Washington in 1955 to affiliate with the Harvard College Observatory. The affiliation was strengthened and formalized in 1973 by the creation of the Harvard Smithsonian Center for Astrophysics under a single director with a joint appointment to SI and Harvard University. Although established mainly as a center for solar studies, over a century later SAO is a research center active in nearly every kind of astronomical observation and at nearly every wavelength, from the gamma-ray regime to the radio regime—a characteristic that SAO shares with no other observatory in the world.

The Zoo began as a collection of live animals used as taxidermists' models. The collection soon became a sufficiently popular public attraction that Congress created NZP and placed it under the direction of SI. NZP was officially opened to the public in 1891. In 1975, a center for the conservation-related activities of NZP, called the Conservation and Research Center, was established in Front Royal, Virginia, to encourage the advancement of the conservation of biological diversity.

The history of the Tropical Research Institute dates back to 1923 when Barro Colorado Island, which was created by the construction of the

BOX 1-1 Timeline of Key Historical Smithsonian Institution Events

1846	**Smithsonian act of organization enacted by Congress**
	President James K. Polk signs Smithsonian act of organization into law
1848	Smithsonian publishes its first book, *Smithsonian Contributions to Knowledge*
1849	Smithsonian initiates International Exchange Service
1855	Smithsonian building completed
1858	Smithsonian is designated the National Museum of the United States
1879	Congress establishes the Smithsonian's Bureau of Ethnology
1881	Arts and Industries Building opens in October
1890	**Smithsonian Astrophysical Observatory established**
1891	**National Zoological Park opens in April in the Valley of Rock Creek**
1910	**National Museum of Natural History opens to public in March**
1943	Freer Gallery of Art opens
1946	**Smithsonian Tropical Research Institute made part of the Smithsonian**
1963	**Conservation Analytical Laboratory (now Smithsonian Center for Materials Research and Education) established**
1964	National Museum of American History opens in January
1965	**Chesapeake Bay Center for Environmental Studies (now Smithsonian Environmental Research Center) established**
1967	Anacostia Museum opens in September
1968	National Museum of American Art and National Portrait Gallery open in Old Patent Office Building
	Cooper-Hewitt National Design Museum becomes part of the Smithsonian
1972	Renwick Gallery opens in January
	Hirshhorn Museum and Sculpture Garden opens in October
1976	National Air and Space Museum opens in its own facility in July
1978	National Museum of African Art established
1983	Museum Support Center opens in Suitland, Maryland
1987	Arthur M. Sackler Gallery opens in September
1989	National Museum of the American Indian established
1990	National Postal Museum established
1994	National Museum of the American Indian Gustav Heye Center opens in New York City
1999	National Museum of the American Indian Cultural Resources Center opens in Suitland, Maryland

NOTE: Events relevant to the research centers in this study are in boldface type.

Panama Canal, became one of the first biological reserves in the New World. Charles Doolittle Walcott, the fourth Secretary of the Smithsonian, began a major biological survey of the Panama Canal Zone. SI was originally one of several organizations participating in research and administration at Barro Colorado Island, but in 1946 Barro Colorado Island became a unit of SI dedicated to conducting long-term research in tropical biology. In 1966, the organization changed its name to the Smithsonian Tropical Research Institute and expanded the scope of its research by establishing marine science centers on the Atlantic and Pacific coasts of Panama and the geographical range of its research by extending its work to other tropical countries. Its broad research interests were legally recognized by the government of the Republic of Panama in 1974, and the relationship of STRI and the Republic of Panama was formalized in the 1977 Panama Canal Treaties. In 1985, the government of Panama granted the Institute status as an international mission; and in 1997, Panama agreed to extending STRI's custodianship of the facilities beyond the termination of the Panama Canal Treaties. Today, STRI is the oldest tropical research station in continuous use and works not only in Panama but throughout the tropics. The Institute has recently signed a contract with the government of Panama whereby it is authorized to continue its research activities and maintain its management of the Barro Colorado Nature Monument with the status of an international mission for a further 20 years.

The Smithsonian Center for Materials Research and Education has its origins in SI's establishment of the Analytical Laboratory in 1963 "to provide information about the objects in the collections of the Smithsonian Institution that is not available through existing facilities. Ways of obtaining information that is required to describe how an object is made, what materials it is made out of, the state of condition or deterioration, and the conservation treatment to be applied in other than routine cases will be investigated and employed." Although the facility, renamed the Conservation Research Laboratory in 1964 and later the Conservation Analytical Laboratory (CAL), was not charged to perform routine conservation, it became overloaded with such requests. In 1978, the Senate Committee on Rules and Administration instructed the Smithsonian to develop plans for CAL to become solely a center for conservation research and education as part of a new museum support center, stressing that the laboratory was not to perform service work for Smithsonian museums, but rather to focus on research and education that would benefit all museums. In 1998, CAL was renamed the Smithsonian Center for Materials Research and Education to reflect its mission in research on preservation, the technical study and analysis of museum collections and related materials, archaeometry, and the organization of conservation training programs.

The Smithsonian Environmental Research Center is the most recently established of the SI scientific research facilities. Originally called the Chesapeake Bay Center for Environmental Studies (CBCES), SERC was established on 368 acres bequeathed to the Smithsonian by Robert Lee Forest on his death in 1962. In 1965, CBCES was established for the conduct of natural history and ecological research programs, especially on the Chesapeake Bay. On July 1, 1983, the facility was renamed the Smithsonian Environmental Research Center after its merger with the Radiation Biology Laboratory, formerly part of SAO. Over the years, several owners of the neighboring land donated their properties to SERC, and SI purchased more of the surrounding property. Today, SERC encompasses 2700 acres, including a completely protected watershed of the Rhode River, a subestuary of the Chesapeake Bay, and 12 miles of undeveloped shoreline.

Although the research units arose as a series of historical contingencies owing to circumstances, dominant personalities, or the availability of funds, the Institution has forged the various branches into a powerful force advancing research, education, and outreach to the public. Considered as a whole, the collection of research units is a major stimulus of continuing public awareness and support of science in the United States and constitutes a distinctive and distinguished addition to the federal research establishment. No government institution maintains a research capacity of such breadth. Ranging from molecular to cosmic scales, scientific research at the Smithsonian includes topics of consequence, such as the population genetics that now undergirds conservation of rare and endangered species worldwide, the long-term databases with which the effects of human activities on terrestrial and aquatic ecosystems can be sorted from normal system dynamics, and indeed the universe. With respect to subjects, research methods, temporal and spatial dimensions of the research, relevance to both long-standing and current scientific issues of importance to the nation, modes of operation, funding mechanisms, and means of administration, the research units of SI collectively add diversity to the nation's overall science enterprise.

BUDGET OVERVIEW AND MOTIVATIONS FOR THIS STUDY

Since the inception of the trust, the US government has generously supported the Smithsonian financially. Although the construction of SI's headquarters building ("the Castle") was financed by the interest accrued from the Smithson trust, the federal government shouldered the expenses of moving the collections and of the care of the collections thereafter. For about 30 years, the Department of the Interior (DOI) reimbursed the Smithsonian for those expenses with funds from its own budget. As the annual contribution from the government increased, Congress, the Secretary of the Smithsonian, and the Secretary of the Interior agreed that it would be more efficient for the Institution to receive direct appropriations from the federal government. Hence, SI became a participant in the federal budget process. In FY 2001, SI had a total budget of about $665 million, of which 57% came from direct federal appropriation. The remainder of the budget is supplied from what SI terms "trust funds," which include income from private donations and contributions; research grants and contracts from such sources as the National Science Foundation (NSF), the National Aeronautics and Space Administration (NASA), and private foundations and nonprofit organizations; proceeds from Smithsonian business ventures (shops, magazine, and so on); and investment earnings.

Table 1-1, which originates from the FY 2003 presidential budget document, provides information on the research budgets of the federal agencies that support scientific research. The total research funding appropriation to SI was $108 million and $111 million for FY 2001 and 2002, respectively—the smallest research budget of the organizations listed. Table 1-1 also shows how the listed agencies classified the allocation of their research budgets among categories of varying levels of merit review, as specified by Office of Management and Budget (OMB) Circular A-11. The review categories of Circular A-11 include research performed at congressional direction ("earmarks" not subject to merit review), "inherently unique" research, merit-reviewed research with limited competitive selection, merit-reviewed research with competitive selection and internal (program) evaluation, and merit-reviewed research with competitive selection and external (peer) evaluation. Inherently unique research is defined as "intramural and extramural research programs for which funding is awarded to a single performer or team of performers *without competitive selection*. The award may be based on the provision of unique capabilities, concern for timeliness, or prior record of performance" (emphasis added). The Smithsonian classifies its entire federally appropriated research budget as inherently unique research.[2]

Competitive processes, such as merit-based peer review, are well established means of setting research priorities in the federal agencies that support US science, with the National Institutes of Health (NIH) and NSF being the premier examples. Prospective peer review is widely regarded as a reliable and fair way to support major science programs over a long period, and competitive grant programs have been recommended repeat-

[2] The Committee was not informed of the rationale for this classification by SI, and no information was provided to the Committee about how other agencies apply this term to their own budgets.

TABLE 1-1 Allocation of Federal Research Funding, 2001 and 2002 (Budget Authority, Millions of Dollars)

Agency	Research Performed at Congressional Discretion		Inherently Unique Research		Merit-Reviewed Research with Limited Competitive Selection		Merit-Reviewed Research with Competitive Selection and Internal Evaluation		Merit-Reviewed Research with Competitive Selection and External Evaluation		Total	
	2001	2002	2001	2002	2001	2002	2001	2002	2001	2002	2001	2002
Health and Human Services	89	142	206	230	2392	2718	201	216	17777	20126	20665	23432
Energy	134	223	1078	1068	2382	2820	305	395	821	788	4720	5294
Defense	678	426	295	350	1012	1014	2712	2950	247	221	4944	4961
National Aeronautics and Space Administration	230	287	152	149	532	398	1377	1550	1894	2291	4185	4675
National Science Foundation	0	0	0	0	191	206	184	192	2700	2887	3075	3285
Agriculture	105	122	815	893	720	676	0	0	206	157	1846	1848
Commerce	18	21	354	377	100	108	204	218	142	166	818	890
Veterans Affairs	1	0	0	0	2	2	349	370	381	408	733	780
Interior	27	48	156	154	379	392	26	31	2	3	590	628
Transportation	55	82	69	73	0	0	338	380	0	0	462	535
Environmental Protection Agency	39	60	39	38	195	192	69	68	133	130	475	488
Education	5	0	0	0	0	0	0	0	169	180	174	180
Smithsonian	0	0	108	111	0	0	0	0	0	0	108	111
Other	385	413	11	7	17	17	76	74	6	6	495	517
TOTAL	1766	1824	3283	3450	7922	8543	5841	6444	24478	27363	43290	47624

edly as a way to ensure the high quality of funded research (e.g., National Research Council, 1994, 1995, 2000). Most academic researchers and many federal researchers regularly compete for grants from such agencies as NSF, NASA, the Department of Energy (DOE), the Department of Defense (DOD), and NIH. Ensuring the best and most responsible use of public funds by increasing the proportion of federal science research that is subject to merit review has been a recurrent theme under many administrations. For example, the guidance to agencies for the FY 1996 budget issued by John H. Gibbons, Director of the Office of Science and Technology Policy (OSTP), and Leon Panetta, Director of OMB, stated a Clinton administration policy that "research not subject to merit review with peer evaluation is expected to decline and funding in these areas should be moved into areas of merit reviewed research with peer evaluation" (Gibbons and Panetta, 1994). The administration's current interest in ensuring high levels of merit review for the federal science portfolio is by no means a new concern.

The federal research funding appropriations to SI reported in Table 1-1 were $108 and $111 million in FY 2001 and 2002, respectively. Those numbers include research expenses at the six scientific research centers covered by this study, additional nonscience research carried out in other parts of SI, and overhead and administrative costs (Smithsonian Budget Office, pers. comm. to Evonne Tang, National Research Council). The classification by SI of its appropriated research funds as "inherently unique" implies that these funds are being spent at SI's discretion on projects of its choosing and without competition. The apparent lack of competition could be interpreted to mean that the work of Smithsonian scientists is not subject to the same rigorous evaluation as that of their academic peers, and this in turn might call into question whether such use of public money is producing research of the highest quality.

The federally appropriated research budgets of most of the SI science centers are supplemented by so-called "trust funds," which is actually a catchall term for funds other than those received through direct federal appropriation or transferred from the appropriations of other federal agencies to SI for services. For purposes of this study, trust funds can be divided into two major categories according to their source: (1) government grants and contracts awarded to Smithsonian researchers through competitive processes, and (2) donations, gifts, endowment funds, and business income (Table 1-2). The direct federal appropriations to SI are used largely to cover 12-month salaries and infrastructure costs, and another concern that has been voiced is whether SI scientists have an unfair advantage over researchers in universities and elsewhere who do not receive similar support. Does SI's receipt of federal appropriations somehow distort the "playing field" on which the US scientific research community competes for research funding?

TABLE 1-2 Estimated FY 2001 Research Budget of Six Smithsonian Research Units by Source (Research Budget, Millions of Dollars)

	Federal Funds		Trust Funds		
Unit	Federal Appropriation	Federal - Other[a]	Government Grants and Contracts	Other Trust[b]	Total Research Budget
NMNH	14.8	0.2	1	4.8	20.8
SAO	24.9[c]	0.4	54.7	3.9	83.9
NZP[d]	3.4	0.2	0.6	2.1	6.3
STRI	6.1	0.3	1.2	2.2	9.8
SCMRE	1.2	0.2	0	0	1.4
SERC	2.1	0.1	2.2	0.8	5.2
TOTAL	52.5	1.6	59.7	13.8	127.4

NOTE: See the companion report by NAPA (2002) for a more detailed discussion of funding for research.

[a]Transferred from other federal agencies.

[b]Portion of endowment income, business income, and gifts raised by the research centers or allocated to them by SI.

[c]Includes $7 million in a separate appropriation for the construction of such major scientific instrumentation as the multiple-mirror telescope and submillimeter array.

[d]Because NZP reclassified its projected expenses for FY 2002 to reclassify some staff as collection staff rather than research staff, the estimated expenses for FY 2001 shown in this table will not match those shown in the NAPA report (2002). The NAPA report uses the new expense classification for FY 2001, FY 2002 and FY 2003.

In the administration's FY 2003 budget document, OMB suggested the commissioning of a study

> to recommend how much of the funds directly appropriated to the Smithsonian for scientific research should be awarded competitively. The review will encompass all Smithsonian scientific research. It will focus on enabling Smithsonian scientific research to compete on a level playing field with other potential performers of the research, where that potential exists. Following the review, if appropriate, the Administration will submit its request to transfer necessary amounts from the Smithsonian to the National Science Foundation.

SCOPE AND EXECUTION OF THIS STUDY

After the release of the President's FY 2003 budget, a process involving contact between OMB and the Smithsonian led to the formulation of a charge for two parallel studies to be conducted by the National Academy of Sciences (NAS) and the National Academy of Public Administration (NAPA). In response to the charge presented to NAS by the Smithsonian, the National Research Council appointed the Committee on Smithsonian Scientific Research to conduct the review with the following questions as its charge:

1. *Are there portions of the Smithsonian research portfolio, which for reasons of their special contribution or uniqueness, should be exempted from being prioritized within that field via a competitive peer reviewed grants program open to all researchers in the public and private sector? Conversely, could some or all of the funds now allocated by the federal government as support for Smithsonian science programs be used more effectively for science if the funds were awarded through a competitive process open to all research performers?*
2. *What are the implications for Smithsonian science programs and for the relevant scientific fields if only those Smithsonian science programs determined to be unique or exempt continue to receive direct federal appropriations?*
3. *For those exempted Smithsonian science programs, how should the quality of this work be regularly evaluated and compared against other research in the relevant fields?*

The Committee was asked to apply that charge to the six scientific research centers under the management of the SI Under Secretary for Science—NMNH, SAO, NZP, STRI, SCMRE, and SERC. The Committee was not asked to address the research centers that report to the Under Secretary for American Museums and National Programs. Nor was it asked to assess the quality of research *per se* at the six centers. Those issues are being evaluated by the Smithsonian Science Commission, which is expected to deliver its report to the Board of Regents at the end of 2002.

The 13 members of the Committee on Smithsonian Scientific Research were chosen for their expertise in the fields of research conducted by the SI science centers covered by the study (astrophysics, ecology, tropical biology, marine biology, biogeochemistry, environmental science, anthropology, paleontology, volcanology, and the collection and preservation of museum specimens) and, where possible, their knowledge of the science and understanding of the roles of the six SI science centers in the broader scientific community. The Committee membership also includes museum directors and academic scientists with extensive relevant experience in institutional management. (The biographies of the Committee members may be found in Appendix A.)

The Committee held its first meeting on May 28-29, 2002, to gather information on SI and its research centers and to hear from representatives of OMB and OSTP. It also heard presentations on how DOE, NASA, and NIH allocate their research budgets through open competitive and other processes. (Although a speaker from NSF was not able to participate in this meeting, committee members and staff interviewed a number of NSF staff during the study.) To facilitate its work, the Committee divided into three panels—on astrophysics (to address SAO), on ecology, environmental science and conservation (to address NZP, STRI, and SERC), and on museum and materials research (to address NMNH and SCMRE). Each panel met with facility directors and other research unit representatives for data-gathering and discussion. After the first meeting, the panels and the Committee's executive group (composed of the Committee chair and the panel leaders) met often by teleconference to draft the Committee's report. The Committee met for a second and final time on July 30-31, 2002, to discuss its findings, reach consensus on its recommendations, and agree on the final report. Throughout the process, the Committee also kept in close contact with the staff and committee that were conducting the parallel NAPA study. (See Appendix B for the NAPA panel's charge and membership).

Chapter 2 of this report describes the SI scientific research units, including their funding structure, research, and outreach activities. Chapter 3 contains the Committee's overall findings, conclusions, and recommendations. Although each panel developed text for this report relevant to the centers it examined, this report constitutes a consensus of the full Committee as agreed to at its final meeting.

2

Description of the Smithsonian Scientific Research Centers

This chapter describes each scientific research unit covered by this study. The descriptions include background information on the centers, information on their budgets, an analysis of how the terms *uniqueness* and *special contribution* may apply to each unit, and other information that the Committee considered relevant.

INTERPRETATION OF TERMS IN THE CHARGE

The charge to the Committee was to determine whether any part of the Smithsonian research portfolio should be exempt from open competition for federal support. The Committee established a framework of criteria to apply to its review of the Smithsonian research centers in the execution of its task. The Committee used the following set of criteria as a guideline; it is not an exhaustive list of all the factors considered.

A. The nature of the Smithsonian as a scientific institution, including

- The role of the Smithsonian in the nation's research complex.
- The interplay between the six research centers.
- The research, education, and public outreach activities of the centers.
- How the loss of one or more of the centers from the Smithsonian would affect the Institution as a whole.

B. How uniqueness and special contribution apply to each of the six research centers. In the context of this study, the terms are complex and may have many meanings that refer to special attributes of a particular research center, such as

- The terms of the center's creation.
- The location and ownership of, or agreements on, property in use by the center.
- The scope of the science engaged in and how it is related to the current status of the center.
- The existence, if any, of one-of-a-kind datasets that confer special significance on the research carried out by the center.
- The ties between the research carried out by the center and the collections of the Smithsonian Institution.

C. How opening some of or all the support now given to each of the centers to a competitive process would affect the science involved. Questions considered included:

- Would the cyclic nature of a competitive process have a favorable or adverse effect on achieving the goals of the field?
- How would opening the research to a competitive process affect the resources of the research centers, including personnel?
- Would any detrimental effect be outweighed by the enhancement of opportunity that a competitive system would give to relevant fields to reach their scientific goals?

D. Given the assumption that any of the six research centers are deemed to be unique and to warrant retention of their current system of support, what recommendations can be made for regular evaluation of the centers to ensure that the quality of their science is maintained? The Committee considered

- The evaluation structures and methods currently in place.
- How these evaluation schemes compare with systems in place in similar institutions.

In addition, in keeping with its charge to address whether any portions of the Smithsonian research portfolio should be exempt from priority setting through a competitive peer-reviewed grants program, the Committee considered the current role of competition in the funding of Smithsonian research.

NATIONAL MUSEUM OF NATURAL HISTORY

NMNH is the largest of the Smithsonian Institution's museums and is the most visited natural history museum in the world. Established in 1910, NMNH was the first Smithsonian unit to be housed in a building constructed specifically for its collections, research facilities, and exhibits. NMNH is SI's largest research center and employs a staff of over 550, of whom about 140 are federally funded research staff, including about 100 curators with PhDs. The current number of curators is about 25% less than the number 10 years ago.

Research at NMNH provides information vital to understanding the dynamic geological, biological, and cultural patterns and processes that have shaped our world. The research center comprises the Departments of Anthropology, Mineral Sciences, Paleobiology, and Systematic Biology. Each department pursues two kinds of intertwined activities: basic research and public educational outreach. The value and appeal of the Museum's exhibits are derived as much from its staff members' scientific expertise and interpretations as from the unique quality of its collections.

The research of the Department of Anthropology addresses three interconnected themes: human interaction with the natural environment, human biology and cultural processes, and human communities in a changing world. Even though their main geographic focus is on North America, the research programs cultivate a broad outlook that contributes to intercultural understanding and enhances the comprehension of humankind's role in the processes of global change. In the past, innovative members of the department originated the distinct field of museum anthropology; today, they provide assistance to law enforcement agencies by sharing their expertise in forensic biology.

The Department of Mineral Sciences is dedicated to understanding the origin and evolution of the Solar System, Earth processes and their products, and the effects of geologic and meteoritic phenomena on Earth's atmosphere and biosphere. As the only physical science department in the museum, it is responsible for maintaining a large and expensive suite of analytical instruments. The group is responsible for some of the museum's best-known exhibits on subjects that include gems (such as the Hope Diamond), volcanoes, and meteorites. These popular exhibits typify the extensive public outreach activities carried out by the curators and other staff scientists in the department.

The Department of Paleobiology houses NMNH's collections of millions of fossil plants and animals, including the popular dinosaur collections, and geologic specimens (rock and sediment cores and samples). Among the topics investigated by members of the department are how long-term physical changes in ancient global geography and climate have affected the evolution of plants and animals, how ecosystems have responded to the changes, and how these responses have influenced today's patterns of biodiversity. The department's studies also provide insights into the processes that control the evolution of species and ancient ecosystems.

Scientists in the Department of Systematic Biology use both time-honored and advanced techniques in phylogenetics, computer analysis, and biochemical comparison to describe and name some of the millions of species that remain unclassified and to understand their relationships in the tree of life, whose branches link all organisms. The department's taxonomic coverage ranges from microorganisms to mammals and from protists to plants and includes the wide variety of terrestrial, freshwater, and marine life. In addition to the breadth of organisms studied, its geographic scope is global. At a time when there is a greater demand for biodiversity information than ever, the NMNH collections are an invaluable source of required knowledge. The work of the museum further develops our understanding of species richness and of habitat degradation. In a unique interagency cooperation, the NMNH curators are joined by 40 systematists from the US Department of Agriculture (USDA), the Department of Commerce (DOC), DOD, and DOI, which colocate their systematics researchers and identification services at the museum to take further advantage of the collections and minimize duplication of effort. Those partnerships contribute $6.5 million dollars per year, an important source of cost-sharing for NMNH (Miller, 2001).

Research Support[1] and Research Output

Of the $46.7 million of direct federal appropriations and transfers to NMNH for FY 2001, $15 million (32%) was allocated to research, and the remainder was used to cover expenses for collections and administrative and other infrastructure costs (Table 2-1). In fact, over 80% of the federal research appropriation was allocated for payment of $12.4 million for salary and benefit costs of research personnel in FY 2001. It should be noted that this $12.4 million for "research" covers the salaries of about 100 curators and other staff who perform many duties—including maintenance and care of collections, exhibitions, and educational outreach—in addition to research.

According to figures provided by SI, government grants and contracts supplemented the research budget at the level of about $1 million per year. Sources of government contracts received by NMNH curators in FY 2001 were NASA, NSF,[2] DOC, DOD, DOI, USDA, and the US Fish and

[1]Figures on research budgets were provided by SI.

[2]Past guidance (NSF Circular 108) stated a policy that SI research staff who are federal employees are ineligible to apply for NSF research funds. Although this policy is apparently no longer in NSF's current grant policy manual, it is apparently still adhered to by many NSF program managers, so that the application of this rule is applied unevenly across NSF Directorates. In addition, some SI staff have been successful in securing NSF support through collaborative projects with other institutions in which they act as coprincipal investigators.

TABLE 2-1 Estimated Research Expenses of NMNH by Source for FY 2001 (in Millions of Dollars)

	Federal Appropriations	Federal—Other[a]	Government Grants and Contracts	Other Trust[b]	Total
Research	14.8	0.2	1.0	4.8	20.8
Salaries and benefits	84%	0%	9%	24%	66%
Other research costs[c]	16%	100%	91%	76%	34%
Other expenses[d]	31.9	1.0	0.8	11.7	45.4
TOTAL	46.7	1.2	1.8	16.5	66.2

[a]Represents appropriations transfers from federal agencies.

[b]Represents portion of endowment income, business income, and gifts raised by the research centers or allocated to them by SI.

[c]Includes travel and transportation; rent, communication, and utilities; printing and reproduction; other contractual services; supplies and materials; equipment and structures; and other costs.

[d]Includes expenses for collections, exhibits and education, administration, facilities, and security and safety.

Wildlife Service (USFWS). Funds were also received from state and regional governments, nonprofit organizations, and private foundations. Only $400,000 of government grant and contract funds and $3.5 million of trust or gift and endowment funds were applied to salary costs.

Because SI receives a direct federal appropriation, its federal employees are not eligible for NSF support as a general policy, except in special circumstances when their contributions are deemed unique. Even with the latter exceptions, opportunities for Smithsonian curators are more constrained than those investigators in other eligible institutions.[3] (Of all federal science agencies, NSF supports research that is, in general, most closely related to the topics covered by NMNH.) During discussions with NMNH department heads, the Committee found that there was considerable variation among the Museum's departments in terms of success in winning research funds from NSF. The Department of Anthropology, for example, reported that it does not have funding from NSF and is barred from applying for NSF grants. The Department of Paleobiology reported, however, that it has had considerable success in obtaining funding from

[3]The Committee consulted the NSF program directors for biotic surveys and inventories, and systematic biology regarding the eligibility of SI curators to apply for NSF grants.

NSF. The Department of Systematic Biology reported that it has obtained funding from NSF only when its scientists have partnered with university researchers and the names of SI scientists have not appeared on the grant applications. Adherence to the policy on funding research at the Smithsonian appears to vary across NSF; some parts of the foundation are more willing than others to entertain proposals from SI scientists. This explains partly why only about 30 of 102 curators in recent years are listed as coprincipal investigators on NSF grants. (According to data provided by NAPA, only 27 of 259 federal grants awarded to SI researchers in FY 2001 came from NSF. NASA made 195, the most from any agency. Sixty-three grants were also awarded by non-federal sources. See NAPA 2002.)

The NMNH curators published 1328 articles during 1995-2000. Journals in which the articles were published include *Nature, Science, Ecology, Oecologia, Proceedings of the Royal Society, Systematic Biology, Paleobiology, Journal of Paleontology, the Biological Journal of the Linnaean Society, Zoological Journal of the Linnaean Society, Geology*, and *Evolution.* NMNH curators also contributed to 264 books during that period.

After recruitment into permanent positions, curators are evaluated on the basis of accomplishments by a peer-review system at 3-7 year intervals depending on their seniority. In a procedure similar to that in universities, performance is evaluated by 10 external reviewers, three nominated by the candidate and seven appointed by the department chair. Reviews follow the Smithsonianwide Performance Accomplishment Evaluation Committee (PAEC) procedures. In 1999, NMNH also underwent an unprecedented process of external programmatic review, when NMNH departments were reviewed by three committees of external scientists, resulting in three independent reports. In 2000, an integrating review committee extracted the overarching themes in the three reports and recommended an action plan for NMNH as a whole.

Unique Characteristics and Special Contribution

Of the 142 million specimens and objects in SI's collections, 90% belong to NMNH, and these collections are a central focus of the research performed at the museum. They include the National Gem and Mineral Collection; the US National Meteorite Collection; the US National Herbarium Collection; 40-50 million fossils, plants, animals, and geologic specimens; and 1500 cataloged specimens of dinosaurs. The collections serve as reference materials for investigating the processes that have modified Earth and shaped the human environment. The presence of world-class researchers at the museum optimizes the access to and organization of the collections in response to changing research needs in various fields.

The museum's collections are unique assets with intrinsic value. They reflect a legacy of over a century of research, exploration, discovery, curation, conservation, maintenance, database management, and use for scientific publication, education, and exhibit. Moreover, the collections are resources that serve the work of many scientists and scholars who conduct study visits to NMNH or borrow material for extended study. In the year 2000, for example, the biologists at NMNH hosted over 900 scientific visitors for over 10,000 visitor-days, made 1433 outgoing loans totaling 166,695 specimens for research, and made over 50,000 identifications (Miller, 2001). Clearly, the health of the Smithsonian's collection-based science is of vital importance to all such efforts worldwide.

Collection-based research helps us to understand the diversity of cultures and the wonders of their achievement, the geologic forces that shaped the planet, the early evolution of the Solar System (through clues provided by meteorites and other materials), the history of life on Earth, and the rich diversity of the living world. Those issues are important both scientifically and in a societal context. Notwithstanding recent dramatic breakthroughs in genetics, proteomics, neurobiology, astrophysics, and many other fields, science remains challenged by some of the most urgent and important problems of our times—the rampant loss of biodiversity; the global-scale degradation of water, the atmosphere, and soils; and the emergence of resilient and highly adaptive infectious organisms. NMNH in the very size and scope of its staff and collections uniquely provides the foundation for a sense of our place in the history and evolution of life and for our stewardship of Earth's biota. NMNH researchers draw directly from the grandest of all biological experiments—the evolution of the biota on a 3.6-billion-year scale—to extract insights bearing on our understanding of life patterns, processes, and history and on our application of this knowledge in ways that directly serve human and societal needs. Examples of important research by NMNH scientists include:

- *Global Volcanism Program*: This is the only group in the world dedicated to documenting the current and past activity of all volcanoes on the planet active during the past 10,000 years. Members of the program have analyzed worldwide patterns in volcano-related deaths, examining the different mechanisms by which human fatalities have been caused by volcanoes during historical time. Findings from the program underscore the importance of monitoring and evacuation and contribute to strategic planning.
- *Forensic anthropology*: NMNH researchers are able to distinguish one person's bones from another's by visual examination and DNA analysis. NMNH researchers helped to identify some of the remains of victims of the September 11, 2001, attacks.
- *Applying knowledge of biodiversity to human health*: On the basis of

the evolutionary relationships of yew trees, NMNH researchers identified a species of yew that produces larger quantities of taxol than other species. This work has helped to reduce the production cost of the taxol, a powerful drug used to fight ovarian and breast cancer.

• *Ticks and Lyme disease*: Using museum collections of ticks from the 1940s, NMNH researchers analyzed for the genetic indicators of the bacterium responsible for Lyme disease. They showed that Lyme disease has been present for at least 3 decades, and for much longer at some sites.

• *Mass-extinction events*: NMNH scientists are among the leaders in documenting some of the great biotic catastrophes in the fossil record, including the Permian extinction event of 250 million years ago that may have exterminated more than 90% of the species living at that time. These studies illuminate the nature and pace of extinction and the lag time in recovery of an ecosystem after such mass destruction—patterns instructive for assessing the current wave of biological extinction induced by human activity.

• *Conservation of song birds*: Analyzing diagnostic isotopes in museum collections and feathers collected from living birds, NMNH researchers demonstrated that loss of winter breeding habitat is correlated with a decline in bird populations in their summer habitat. This is an important finding for conservation strategies.

• *Global climate change*: Specimens of deep-water corals at NMNH hold within their mineral parts a record of ocean circulation and global climate for the last 50,000 years. Using the coral collection, NMNH scientists make inferences about global-scale changes that help to inform predictions of future worldwide climate patterns.

• *Understanding the tree of life*: All the world's species are linked in a great branching pattern of relationships that reflects a shared evolutionary history of 3.5 billion years. NMNH scientists bring unique data to bear on understanding the branching relationships of spiders, insects, other invertebrate groups, land plants, birds, fishes, reptiles, mammals, and many other organism groups.

With the museum's important holdings come the responsibilities for care, security, database management, conservation, and access that challenge all large museums. As the holdings become more rare and precious as scientific resources, there are new tools to enhance their access and utility. A recent development that has revolutionized museum collection-based science is the digitization of collection data, supplemented by brilliant digital imagery, remote sensing and geographic information system (GIS) data, and specimen data on size, shape, and other characteristics. NMNH staff members are exploiting these technologies and assembling powerful new databases. Accordingly, NSF has made awards in rare cases to the Smithsonian scientific effort related to unique collections and their database management and imaging.

Other Activities

By statute, NMNH serves the public. Enhancing public understanding of the life sciences, geosciences, and human sciences through exhibits and educational programs is central to its role. One of the best ways to maintain popular interest in the Smithsonian collections is for its exhibits to reflect the latest scientific breakthroughs, and this effort is greatly facilitated by having a curatorial staff that actively participates in leading-edge scientific activities. Indeed, although use of its collections forms the core of its service to the scientific research community, the science performed at NMNH also informs its exhibits and interpretive programs and helps to attract millions of members of the general public as visitors each year. NMNH strives to reach out to audiences of all ages across North America through inhouse and traveling exhibits and supports K-12 science education by providing educational content, teacher training, and student support.

Status of the National Museum of Natural History in the Museum Community

NMNH's mission in research, collection care, database management, exhibition, and public education is shared by such major natural history museums as the Field Museum (Chicago), the American Museum of Natural History (New York City), and the California Academy of Sciences (San Francisco). However, NMNH plays a unique and critical role in the natural history museum community. The collections at NMNH are vastly larger in size and scope than those of any comparable US institution; for example, at 142 million specimens, the NMNH collections are more than 4 times the size of the next largest group of collections (32 million specimens) maintained by the American Museum of Natural History. Its breadth of research mission and the extent of its service to the museum research community are correspondingly greater. The support and function of an institution of the size of NMNH warrant high national priority for collection-based research that is vital to the accomplishments of an international community devoted to the natural sciences.

SMITHSONIAN ASTROPHYSICAL OBSERVATORY

SAO is SI's research center for astronomy and astrophysics. From the earliest days of the Institution, there were those who thought that astronomy should be a central pursuit for an organization charged with "the increase and diffusion of knowledge." In 1890, SAO was established initially mostly as a solar observatory. Over a century later, SAO is a research center with more than 900 employees active in nearly every field of astronomical observation, from the gamma-ray regime to the radio, and with a major science education group.

SAO is in Cambridge, Massachusetts, where it moved from Washington, DC, in 1955 to affiliate with the Harvard College Observatory. The affiliation was strengthened and formalized in 1973 by the creation of the Harvard Smithsonian Center for Astrophysics under a single director with a joint appointment to SI and Harvard University. Of the 907 staff at SAO, 341 had PhDs according to the most recent annual staff census. The research staff is divided among seven research divisions: atomic and molecular physics, high-energy astrophysics, optical and infrared astronomy, planetary sciences, radio and geoastronomy, solar and stellar physics, and theoretical astrophysics. In addition to education and outreach activities in its research divisions, SAO has a department devoted to science education.

SAO owns and operates two major astronomical research facilities: the F.L. Whipple Observatory, including the multiple mirror telescope (MMT), which is operated jointly with the University of Arizona and has been in operation since the 1960s; and the submillimeter array (SMA) on Mauna Kea, under development as a joint project with the Institute of Astronomy and Astrophysics of the Academia Sinica of Taiwan. The conversion of the MMT to a single 6.5-m telescope and the construction of the SMA along with their major instrumentation programs are funded separately from SAO's basic operations and research budget. SAO also has major roles in other astronomy research facilities, the largest of which is operating the Chandra X-ray Observatory (CXO) and its associated Chandra Science Center under contract with NASA. The contract was extended in July 2002 to August 2003.

Research Support and Research Output

As is true of the other SI units, the SAO budget includes funds from direct federal appropriations and other sources. But in contrast with the other units, for which the direct federal appropriation provides the majority of operating funds, the appropriation for SAO represents only about 24% of its total budget. That is because SAO receives substantial funding from federal contracts and grants (59% of its total budget). The largest is a contract with NASA to operate the orbiting CXO. SAO also receives grants from DOE, NSF, the US Air Force, and foreign governments. In FY 2001, 16% of the overall SAO budget was supported by funds from Smithsonian's endowment and business activities and by private funds.

The total research expenditure by SAO in FY 2001 was $83.9 million, of which the federal appropriation was $24.9 million (Table 2-2). The FY 2001 federal appropriation for research also included $7.0 million for major construction of scientific instrumentation (for example, the SMA and MMT). Most SAO staff are paid from trust funds rather than from direct federal appropriations (federal funds and trust funds paid for 64 and 148 full-time-equivalent staff, respectively), and they compete both internally and externally for research support. As noted above, most of SAO's income comes from government contracts and grants obtained through competitive peer review. In FY 2001, SAO was awarded 214 contracts and grants amounting, over the lifetime of the awards, to a total of $88.7 million. In addition, SAO staff compete internally for research project support from trust funds and for access to the observing facilities supported by the federal appropriation. In 1995-2000, an estimated 219 SAO scientists produced 2409 research publications.

SAO has instituted a system of assessment and review of the individual scholarly accomplishments of the staff in addition to the normal annual review required of all Smithsonian employees, both federal and trust-fund. The members of the PAEC are appointed by the SAO director on the advice of the associate director of each division. The PAEC is charged to review the work of each member of the staff above grade G-13

TABLE 2-2 Estimated Research Expenses of SAO by Source for FY 2001 (in Millions of Dollars)

	Federal Appropriations	Federal—Other	Government Grants and Contracts	Other Trust	Total
Research	17.9	0.4	54.7	3.9	83.9
Salaries and benefits	63%	0%	41%	34%	42%
Other research costs	37%	100%	59%	66%	58%
Major scientific instrumentation[a]	7.0	0	0	0	0
Other expenses[b]	2.5	0	12	13.9	28.4
TOTAL	27.4	0.4	66.7	17.8	112.3

[a]In addition to the federal appropriations for research and infrastructure, SAO received $7 million for the construction of major scientific instrumentation, such as the multiple-mirror-telescope and submillimeter array.

[b]Includes expenses for exhibits and education, administration, and facilities.

for trust employees and grade G-12 for federal employees at least once every 5 years. In accordance with SI guidelines on PAEC operation, the review committee does not make specific recommendations to the director with regard to grade level or staff assignment, but it does provide analysis and conclusions as to the professional standing and level of accomplishment of each scientist. If there is a question of demotion or promotion regarding a given scientist, the PAEC solicits five or six letters of evaluation from non-SAO research peers.

An SAO visiting committee reports to the Under Secretary for Science at the Smithsonian. This committee of eight members—drawn on the whole from the wider astronomy and astrophysics community—meets for 3 days every 18 months to assess the health and operation of SAO. Given the impossibility of a detailed review of the entire center by this modest committee in such a short time, SAO is planning to institute a new system of more in-depth divisional review by a panel consisting of internal and external members. The center's computational facilities will be the topic of the first such review. The panel will include one member from outside SAO. SAO plans to hold separate reviews of each of its scientific divisions, again with outside membership.

Unique Characteristics and Special Contributions

A number of major scientific results and/or major observing capabilities have originated in SAO. These have been enabled or facilitated by the presence of a stable core of senior science staff supported by the direct federal appropriation and by the access of SAO staff to the combination of internally peer-reviewed SI resources and externally peer-reviewed grants and contracts. Many of the SAO facilities developed over the decades have been ground-breaking, innovative, and high-risk endeavors that have substantially advanced astronomy and astrophysics. (See Appendix C for a list of SAO facilities). These are examples of important SAO activities:

- The SI-supported *red-shift survey*, the first large-scale spectroscopic survey, which produced the first significant three-dimensional map of a portion of the sky and allowed the study of the large-scale structure of the universe in space and time.
- The successful and still operating *Chandra X-ray Observatory*, for which the concept, development of enabling technologies, and final proposal were developed by members of the x-ray group in the SAO High Energy Astrophysics Division and whose principal investigators and many key personnel were SAO staff members for many years before the advent of dedicated NASA funding.
- The pioneering work in very-high-resolution radio observations–

very-long-baseline-interferometry (VLBI)—including studies of cosmic masers and their recent application to the determination of the cosmic distance scale and direct evidence of the existence of massive black holes in the nuclei of galaxies.

- The early promotion of *direct gamma-ray astronomy*, now a well-established astronomy discipline, and the *search for Cerenkov radiation* from very-high-energy gamma rays, which began as an SAO program in 1968 and led to the first detection of a galactic source of very-high-energy gamma rays in 1989 and the first detection of an extragalactic source in 1992.
- The development of the *submillimeter wave astronomy satellite* (SWAS), an orbiting telescope studying star-forming clouds in the Milky Way. SWAS, a NASA MIDEX project, was awarded to SAO after competitive review.
- The development of the *multiple mirror telescope* (MMT), with advances in telescope design that have been incorporated into almost every new-generation telescope.
- The development of the *submillimeter array* (SMA), under construction.
- The development of the *infrared-optical telescope array*, a three-telescope stellar interferometer that has led to many innovations, including the use of single-mode optical fibers.

The suite of SAO accomplishments has been made possible by steady support from direct federal appropriations, and many would not have been possible if the programs had to depend on the 3-year funding cycles of such agencies as NASA and NSF, even if NSF funding were made available to all SAO staff.

In addition to developing, designing, and constructing facilities that enable SAO scientists to carry out research in nearly every wavelength band of astronomical observation (a characteristic that is peculiar to SAO in contrast with every other observatory in the world), some fraction of the time on SAO observation facilities is made available for external use as a contribution to the well-being and work of the wider astronomy community in the United States and abroad. For example, at the MMT facility in 2001, 81 coprincipal investigators and coinvestigators were involved in 92 research programs. Overall, about 6% of the MMT's operating time is given over to non-SAO users. It is expected that about 10% of the time on SMA will be available for use by astronomers who are not members of the consortium building the array. All observing time on the CXO is awarded by peer review coordinated by SAO's Chandra Science Center under contract with NASA. Typically about 170 non-SAO researchers from about 120 institutions in the United States and abroad are actively engaged in research on SAO facilities.

Other Activities

SAO participates in the education of K-12, college, and graduate students. Unique opportunities are offered to undergraduate and graduate students in pursuit of an education in astronomy. In SAO's Education Department, about 50 staff, two of whom are federal employees, are involved in curriculum development, teacher training, and even television production. There is important synergy between this group and the active SAO researchers. The Chandra project also carries out substantial educational programs as part of its NASA contract. The SAO education and outreach activities contribute substantially to the effort to increase scientific and technical literacy in the United States.

NATIONAL ZOOLOGICAL PARK

As zoos have evolved over the last 25 years, it has been increasingly clear that they have an important role to play in conservation and research, which are now incorporated into most institutional mission statements. Research has been part of the institutional mission of NZP since its inception in 1889 and this role expanded with the founding of the Conservation and Research Center (CRC) in 1974. CRC also serves as a captive breeding center for NZP. The CRC facilities consist of 87 buildings on 3200 acres in Front Royal, Virginia, about 75 miles from NZP. A total of 52 persons staff the facility, of whom 31 are PhD-level staff scientists. CRC research covers a number of disciplines, including reproductive biology, veterinary medicine, conservation biology, species recovery, genetics and genome resource banking, and GIS spatial analysis for conservation.

The CRC science programs, number of staff scientists, and disciplines represented are comparable with those of other zoos of similar size and similar status in the zoo and conservation community, such as the Brookfield Zoo in Chicago (25 scientists), the Institute of Zoology at the London Zoo (22 scientists), and the San Diego Zoo (35 scientists). Larger institutions, such as the Wildlife Conservation Society (Bronx Zoo), have as many as 75 scientists. CRC's management of researchers and research programs is also comparable with that in other major zoological institutions and universities. Hiring practices are open and competitive, and external review of researchers occurs every 3-5 years (recently changed from every 5-7 years). External review of the research programs has taken place less often, but has occurred twice in the last 10 years (1993 and 2001). Review committees examined research and other activities at CRC and considered the future potential of the programs.

Research at CRC is directed toward support of the animal collection and captive breeding programs of NZP. The proportion of effort directed toward support of the animal collection varies with the research discipline and among individual scientists. This is important for two reasons: it makes a large proportion of CRC research inextricable from the mission of NZP, and to the extent that some collection-driven research and support activities are less fundable and less publishable than pure hypothesis-driven research, it has the potential to reduce the likelihood of extramural support and the publication output of investigators. Thus, federal support is required for this essential activity.

Research Support and Research Output

The total expenditures in FY 2001 of NZP amounted to $27.3 million, of which $6.3 million was spent on research. About 55% of the research budget of CRC comes from direct federal appropriations; extramural grants, contracts, gifts, and other funds make up the rest (Table 2-3). In FY 2001, 72% of the total federal research appropriation supported staff salaries and benefits. Nineteen of 27 staff scientists are supported by the federal appropriations, and the rest are paid by trust funds. Most research personnel perform functions in addition to research, such as animal care and health support, education, and outreach. In FY 2001, NZP classified its staff salary on the basis of their predominant job functions; the expenditures on research staff salaries and benefits in the NZP budget do not correlate with the proportion of time that staff spent in performing research. In FY 2002, NZP reclassified its staff salaries to reflect better the amount of time spent in different job functions. The reclassification resulted in a projected budget that is smaller than the FY 2001 budget (see NAPA 2002).

In FY 2001, the research support from sources other than the direct federal appropriation for CRC was $2.7 million. Funding sources included NSF, NIH, DOD, the Environmental Protection Agency (EPA), and private foundations, such as the International Rhino Foundation and the Morris Animal Foundation. The majority of the federal appropriation for research goes to salaries (72%), and competitive extramural grants, contracts, and gifts cover the bulk of other research costs. The situation is comparable to that in other major zoological institutions and universities in which direct support of some kind is used to create the basic research framework, which is then leveraged to generate extramural support for the actual research endeavors.

In 1995-2000, CRC staff scientists generated 643 publications in a broad array of journals, including *Science*, *Nature*, *Conservation Biology*, the *Journal of Reproduction and Fertility*, *Veterinary Pathology*, and the *Journal of Zoology*.

TABLE 2.3 Estimated Research Expenses of NZP by Source for FY 2001 (in Millions of Dollars)

	Federal Appropriations	Federal—Other	Government Grants and Contracts	Other Trust	Total
Research	3.4	0.2	0.6	2.1	6.3
Salaries and benefits	72%	0%	37%	27%	51%
Other research costs	28%	100%	63%	73%	49%
Other expenses[a]	17.7	0	0.1	3.2	21
TOTAL	21.1	0.2	0.7	5.3	27.3

NOTE: NZP reclassified its projected expenses in FY 2002 because some staff who were previously characterized as research staff are better described as collection staff. Therefore, the projected expenses for research in FY 2002 are substantially lower than the estimated expenses for research in FY 2001.

[a]Includes expenses for collections, exhibits and education, administration, facilities, and security and safety.

Unique Characteristics and Special Contributions

The mission of NZP is "to celebrate, study, and protect the diversity of animals." Research is thus an essential element of the NZP mission. In addition, the institution is accredited by the American Zoo and Aquarium Association (AZA), and it is implicit in its mission that the animal collection receive a high standard of care and be used optimally for conservation, education, and research (American Zoo and Aquarium Association, 2002; or see guidelines for accreditation at www.aza.org). Accreditation by AZA is a fundamental prerequisite for zoos because only AZA member institutions are allowed to exchange animals and participate in cooperative breeding programs with other members. This is essential for NZP to maintain self-sustaining animal populations in captivity. Although it is not required for the maintenance of accreditation, the research performed at NZP is consistent with AZA standards, and reinforces the credibility of NZP.

The scientific disciplines represented at NZP and CRC (such as reproductive physiology, conservation biology, and animal health) and the scope of research conducted reflect the close relationship between the animal collection, research activities, and mission of the institution. The research activities also facilitate AZA accreditation and USFWS requirements for some species in the collection; for example, without the NZP's program of high-quality, multidisciplinary research, permits to import giant pandas could not have been obtained. Not all research and conservation activities are directly related to the Zoo's animal collection, however. For example, NZP is engaged in conservation of animals in their native habitats—an important element of the Zoo's special contribution to international efforts in conservation biology, including the hosting of on-site training for conservation specialists from developing nations.

These are important elements of NZP's uniqueness and special contributions:

- The proximity of the animal collections to NZP's and CRC's animal health, pathology, and reproductive physiology researchers, which facilitates rapid response to urgent collection needs.
- Connections to other Smithsonian research centers, including the Smithsonian Institution/Man and the Biosphere Biological Diversity Program (SI/MAB). SI/MAB provides training in biological diversity monitoring. The arrangements between NZP and NMNH, whereby all animals that die at NZP and CRC receive a postmortem examination, facilitate the collection of a variety of research samples and the deposition in NMNH of specimens that would otherwise be difficult or impossible to obtain for the NMNH collections.
- External connections, including partnerships with other zoos in captive-breeding and reintroduction programs (as in the case of giant pandas and golden lion tamarins), collaboration with university-based researchers, and cooperation with the many species survival plans, taxon advisory groups, and the Conservation Breeding Specialist Group of the International Union for the Conservation of Nature. CRC scientists hold adjunct appointments at 20 universities around the world.
- The development of unique and irreplaceable databases, including the archive of the pathology department, which consists of tissue samples and data obtained from animals in the NZP collection over more than 30 years.
- A substantial impact on the understanding of diseases of captive wildlife. The discovery by NZP pathologists of a fungal skin disease in frogs, which has turned out to be a major factor in the global decline of amphibians, is a good example of the pre-eminent work being done.
- CRC's international reputation for excellent work in assisted reproduction, cryopreservation, and endocrinology. The pioneering work of CRC scientists in population genetics and small population management has defined the standards and methods by which all other major institutions manage their breeding populations.

That the research, conservation, and education activities are intrinsically related to the animal collection itself is a key element of the uniqueness of NZP and CRC.

Other Activities

Education, training, and outreach are important at CRC. The Education Office focuses on K-12 education, providing teacher training workshops, resources for students and teachers, and outreach programs to local schools and communities and educational opportunities for undergraduate college students. CRC also has become one of the leading institutions in the world in providing training in wildlife management and conservation for researchers and wildlife managers in developing nations. Such innovative programs are made possible by the conservation and research expertise of CRC scientists combined with facilities that include on-site dormitories and a conference center. A number of additional education and outreach activities are conducted through other CRC programs, such as the Migratory Bird Center, the Conservation Genetics program, the Environmental Latino Initiative Promoting Science Education (ELIPSE), and the GIS spatial analysis laboratory. The number of students and visiting scientists in those programs was not reported, but more than 1200 scientists have gone through the SI/MAB program.

SMITHSONIAN TROPICAL RESEARCH INSTITUTE

STRI is the only federally funded US tropical research station outside the United States and its territories. The Panama-based STRI complex is headquartered in Panama City (1500 ha). Since its establishment in 1946, STRI has expanded into a complex of nine research stations throughout Panama, including its marine biology facilities. STRI is one of three Smithsonian centers that function, in essence, like an academic department with research, educational, and outreach obligations. The core staff of 33 senior scientists with PhDs, most with joint appointments in US and Latin American universities, are recruited internationally and serve as collaborators with visiting scientists. Long-term studies and tropical research projects are being conducted throughout the isthmus at terrestrial and marine field stations equipped with modern laboratories and dormitory facilities. Research programs are in tropical biology, archeology, behavioral ecology, sociocultural anthropology, environmental monitoring, tropical forest ecology, paleoecology, molecular evolution, plant physiology, and tropical marine ecology.

Each year, STRI hosts some 600 visiting scientists and students from some 200 institutions. The Earl S. Tupper Tropical Sciences Library is one of the most comprehensive resources in the world for tropical biology and

TABLE 2-4 Estimated Research Expenses of STRI by Source for FY 2001 (in Millions of Dollars)

	Federal Appropriations	Federal—Other	Government Grants and Contracts	Other Trust	Total
Research	6.1	0.3	1.2	2.2	9.8
Salaries and benefits	74%	0.6%	11%	35%	55%
Other research costs	26%	99.4%	89%	65%	45%
Other expenses[a]	4.9	0	0.2	1.3	6.4
TOTAL	11	0.3	1.4	3.5	16.2

[a]Includes expenses for exhibits and education, administration, facilities, and security and safety.

conservation. The mission of STRI is to "further our understanding of tropical nature and its significance to the world at large, to train students in tropical research, and to promote conservation by making the public aware of the beauty, importance and fragility of the ecosystems of the tropics." A major focus of STRI is long-term studies of tropical biology.

Research Support and Research Output

During FY 2001, STRI received an $11 million federal appropriation, of which $6.1 million was allocated to research. The remaining appropriated funds were expended on education and exhibits, administration, facilities, and security (Figure 2–4). Of the federal appropriation for research, 74% was spent on salaries and benefits. An equivalent of 27 of the 35 senior scientists' salaries[4] is paid by federal appropriations. The research budget was substantially augmented by an additional $3.4 million from extramural grants and contracts, gifts, and other trust funds. STRI research staff were awarded competitive grants by NSF, NIH, NASA, DOD, the US Agency for International Development (USAID), and private foundations, such as Andrew W. Mellon Foundation, the American Cocoa Research Institute, the Turner Foundation, and the International Plant Genetic Resources Institute. Additional research support comes from overseas grants and grants that are awarded to individual scientists instead of to the institution.

[4] Some senior scientists at STRI work part-time.

In 1995-2000, STRI staff scientists generated 511 scholarly publications in a broad array of books and journals—*Science, Nature, Ecology, Trends in Ecology and Evolution, American Naturalist, Proceedings of the National Academy of Sciences*, and other high-impact journals with wide readership and international recognition.

STRI has benefited from the use of peer review committees in evaluating the accomplishments of its professional employees and its programs. External review committees generally consist of four or five individuals with experience and reputation appropriate to senior scientists. Professional evaluation is based on research, dissemination of ideas (including publication of research articles and books, invited and contributed papers at scientific meetings, teaching and educational materials, supervision of graduate students, editorial and review work for scientific journals, and organization of symposia, conferences, and international meetings), and science administration at STRI. Accomplishments of reviewees are assessed by 10 external reviewers, and the results are then examined by the external review committee. Institutional evaluation includes assessment of STRI's research programs, their relationship to other functions of the Institute, and future directions for STRI. STRI was last evaluated by an external visiting committee in 2000.

Unique Characteristics and Special Contributions

STRI is internationally recognized for its outstanding and diverse research on the New World tropics, including the fields of tropical biodiversity and systematics, plant-animal interactions, archeology, behavioral ecology, cultural anthropology, environmental monitoring, tropical forest ecology, paleoecology, molecular evolution, plant physiology, and tropical marine ecology. STRI is one of the few tropical research stations that takes an integrated approach to studying plant and animal ecological interactions. The US Forest Service has stations in Puerto Rico and Hawaii that focus on forests and forest disturbance, but none of these stations performs work with the breadth of that found at STRI.

Both resident and visiting biologists at STRI have the advantage of living at the site of their field work while having access to cutting-edge facilities and laboratories—a situation rarely found in the tropics. Sophisticated large-scale equipment, such as canopy cranes and GIS facilities, is also available at STRI; this is another factor that provides unique advantages to STRI. The opportunity for scientists to exchange ideas and observations makes this station a "think tank" for understanding complex tropical ecosystems. The breadth of these scientific interactions goes far beyond what is possible in a university department, where few specialists are focused on similar or related research questions.

Highlights of STRI research achievements include the following:

- The first and most extensive long-term study of the population dynamics of trees in tropical forests encompassing more than 3 million individuals in large-scale (typically 50-hectare) plots in more than 14 countries throughout the tropics.
- The study of flower and seed production in 625 species of trees, shrubs, and lianas at weekly intervals for 16 years in order to understand the life histories of these species in intact tropical forests.
- The quantification of larval settlement on coral reefs at monthly intervals for 20 years and monitoring of populations of sea urchins for 15 years to understand the processes that control marine biodiversity. These data make up a few of the long-term records against which effects of global change on coral reefs might be assessed.
- More than 2 decades of study of the life histories of various tropical insect species to understand their role and behavior in intact and disturbed tropical forests.
- The pioneering long-term study of the behavior and population dynamics of tropical lizards, birds, and other vertebrates. Some studies have extended for more than 30 years.
- Building of a repository of over 300,000 mollusk specimens from Central America with extensive geographical, sediment, taxonomic, and biological information; this is the most complete collection of this kind in the world.
- Service as the repository of 50,000 marine fossils in which the evolution of marine biodiversity can be traced.
- Continuous recording of rainfall and temperature data for Barro Colorado Island for more than 80 years and operation of a class A weather station there for 31 years.

Because of long-term support for its research programs, STRI can answer "big picture" questions that require continuity over decades. Research carried out over years or even decades is now recognized as fundamental and vital both to scientific understanding and to society's ability to make informed policy choices, for example, about climate change. Many ecological processes vary over decades, and short-term observations of a few months, seasons, or years provide incomplete and inadequate data for understanding ecosystem behavior. This type of long-term program can be effective only at an established site with stable long-term support. Having a stable source of funding, STRI scientists have the unique ability to examine both terrestrial and marine tropical systems over long timeframes. (See Appendix D for examples of long-term projects at STRI.)

In addition to facilitating long-term research, the stability of the current federal support for STRI helps to underpin the strong 90-year relationship of trust and confidence between STRI and the government of the Republic of Panama. In June 1997, STRI signed an agreement with the government of Panama whereby the Institute is authorized to continue its research activities and maintain the custodianship and management of the Barro Colorado Nature Monument and its international mission status for an additional 20 years. Recently, STRI also signed agreements with the Interoceanic Canal Authority (ARI) to ensure continued use of its current structures, areas, and facilities for 20 years upon termination of the Panama Canal Treaties in 2000.

Other Activities

STRI engages in a variety of educational and training activities. Although STRI is not a degree-granting institution, it hosts undergraduate and graduate students from US and Latin American universities who engage in field work in collaboration with STRI staff scientists. The most important educational activity is the fellowship program, which provides support for graduate students, postdoctoral fellows, and senior fellows. In 2000, 38 fellows and 50 interns were supported at STRI, 31% with institutional funds and the remainder with grants. This is an outstanding contribution to science training, surpassing the scope of fellowship support provided by many educational institutions of comparable size. STRI is one of the few places where US students in tropical biology are trained.

SMITHSONIAN CENTER FOR MATERIALS RESEARCH AND EDUCATION

Originally established as the Analytical Laboratory of the United States National Museum in 1963, SCMRE is a unique organization whose primary purpose is to increase and disseminate scientific knowledge that contributes to improved preservation and conservation of museum collections and related materials and contributes to enhancement of their contextual interpretation.

All museum curators are acutely aware of the need to preserve the materials in their care, as are any museum visitors who have observed the ravages of time on their favorite exhibits. After its founding in 1963, there was some expectation that the center would be a service organization helping to conserve the general collections of the Smithsonian. Its actual role, however, is much broader and more valuable. In its conservation activities, SCMRE seeks to identify and solve general problems in need of solution for the museum community as a whole rather than focusing exclusively on the day-to-day conservation of particular specimens.

Similarly, its research for extracting information from the collections harnesses modern developments in nuclear chemistry, molecular genetics, instrumental analysis, and related fields. Rather than trace the source of a particular pre-Columbian pot or the foundry where a particular Middle Eastern bronze axe was produced, SCMRE develops and then applies new methods, such as neutron activation analysis of ceramic or metallic artifacts, to trace the sources of such materials. Although research and teaching of this sort are carried out to some extent by universities and museums elsewhere, few centers have SCMRE's breadth of purpose or impact.

Research Support and Research Output

In FY 2001, SCMRE scientists received $3.5 million in direct federal appropriations, which made up almost the entire SCMRE budget (Table 2-5). The majority of the appropriated funds are allocated to administrative and infrastructure costs (62%). Of the $1.2 million expended on research, 80% was used to pay staff salaries and benefits. All staff scientists at SCMRE are paid by federal appropriations.

Until the 1980s, SCMRE staff were discouraged from applying for extramural funding. Even now, applying for funding from some federal agencies is generally not successful because no federal agency has a dedicated program in this discipline. To overcome that barrier, SCMRE staff have been collaborating with scientists outside the Smithsonian and receiving some "in kind" support through partnerships. For example, some travel and field expenses and laboratory supplies were provided by the

TABLE 2-5 Estimated Research Expenses of SCMRE by Source for FY 2001 (in Millions of Dollars)

	Federal Appropriations	Federal—Other	Government Grants and Contracts	Other Trust	Total
Research	1.2	0.2	<0.05	<0.05	1.4
Salaries and benefits	80%	0%	88%	0%	69%
Other research costs	20%	100%	0%	100%	31%
Other expenses[a]	2.0	0.1	0	0.1	2.2
TOTAL	3.2	0.3	0	0.1	3.6

[a]Includes expenses for collections, exhibits and education, administration, and facilities.

National Park Service, the National Center for Preservation Technology and Training, NSF, Kress, and the Latino Fund for Scholarly Studies through collaborations. These external funds are not reflected in the SCMRE budget because travel or other support costs were paid directly by external collaborators and there was no money transferred to SI.

SCMRE has a staff of 26, including nine PhD scientists. The research interests of the staff encompass a variety of disciplines, including the application of organic chemistry to museum conservation, ceramic science, materials characterization, chemistry of archeological and organic materials, mechanical properties of modern materials, the effects of environmental influences (e.g., moisture) on these materials, metallurgy, and the conservation of furniture, paintings, paper, and textiles.

Thirteen staff members published 133 articles in 1995-2000, including 19 contributions to books. In addition to publication in the professional literature, staff members published technical monographs on such subjects as insect pests in museums and historical textile dyestuffs. They also produced technical videos for nonconservation collection professionals.

All staff scientists at SCMRE are evaluated periodically by PAEC. Each staff scientist is required to submit a curriculum vitae, forms describing the reviewee's research and professional and educational service, copies of publications, and letters from outside reviewers to the director, who passes the information on to the PAEC. The PAEC commonly includes scholars from outside the SI and occasionally one non-SCMRE SI scholar.

Programs of the research unit are also reviewed periodically by external visiting committees. Visiting committees generally review the mission statement and accompanying documents and the effectiveness, relevance, and importance of the programs and make recommendations for long-term priorities. SCMRE was last reviewed by a visiting committee in 1995 when it was known as the Conservation Analytical Laboratory.

Uniqueness and Special Contribution

One of the goals of the SCMRE, in its own words, is "to be a driving force for the promotion and improvement of conservation and preservation of Smithsonian collections." SCMRE has a unique and important role in the Smithsonian, as it provides analytical, technical, and information support for the care and curation of institutional collections at the request of Smithsonian staff. Furthermore, SCMRE staff members assist and support the Smithsonian administration in the design and execution of collection care and management. In 1996-2002, SCMRE performed 4470 analyses for various Smithsonian museums, including NMNH, the National Museum of American History, the National Air and Space Museum, the Freer Gallery of Art, the Arthur M. Sackler Gallery, the Hirshhorn Museum and Sculpture Garden, the Smithsonian American Art Museum, the National Museum of African Art, and the National Museum of the American Indian. Apart from SI staff, faculty and students in various universities in the United States, Canada, and Argentina also use the analytical facilities and unique databases that SCMRE offers.

Well recognized achievements and special contributions made to its field of research include the following:

- A 500-fold increase in the stability of storage of 20th century photographic collections through the development of low temperature (–20°C) storage.
- In-depth studies of the aging effect of temperature and humidity in the curation of organic materials.
- Unique research focused on the preservation of the Star Spangled Banner.
- The analysis and reconstruction of ancient technologies, leading to a deeper understanding of such issues as pottery making among native Californian and Mayan cultures.
- The development of advanced artifact packing guidelines based on shock and vibration analysis.
- Special research in archaeometallurgy with such results as the demonstration that the Wright brothers' use of the alloy duralumin was the earliest by 6 years.

The unique role and special contribution of SCMRE are possible only because of the federal support for its unique niche. The specialists it recruits have accumulated experience in the application of their analytical techniques in a museum context, and this experience is vital to the success of the research and the training and display programs.

The long-term research in cultural and archeological materials that SCMRE undertakes is also unique. In partnership with the National Institute of Standards and Technology, SCMRE produces high-precision analytical chemistry data for compositional characterization in studies of archeological, historical, and art history research projects using instrumental neutron activation analysis (INAA). SCMRE maintains all these data in an INAA database that is accessible to its collaborators around the world.

It is unlikely that the types of projects SCMRE excels at would, or could, be supported by a system of competitive grants. First, no science funding agency in the federal government has a mission that overlaps extensively with the needs of the museum community. Second, SCMRE's long-term projects are not compatible with the federal science agencies' typical 2- to 3-year cycles of support.

The work of SCMRE is paralleled by few institutions in North America. Two similar institutions are the Getty Conservation Institute (GCI) and the Canadian Conservation Institute (CCI), both of which specialize in research that differs somewhat from SCMRE's. GCI is a non-profit organization that obtains its funds almost exclusively from the J. Paul Getty Foundation. It conducts research that is related primarily to the conservation of architecture, archeological sites, monuments, and large objects, such as statues and wall paintings. Because it is a private foundation, its research is not necessarily driven by requests from museums. CCI is a special operating agency of the Department of Canadian Heritage in the Canadian government. Its work is mostly supported by direct appropriations from the Canadian government, even though it generates some revenue from contracts with private and international clients. CCI's four main activities are treatment, analysis, research, and transportation. CCI treats and analyzes objects from various museums, galleries, and archives in Canada. Projects range from research on materials used by Canadian artists to work on a standard for permanent paper. CCI also provides transportation services. Like SCMRE, GCI and CCI have stable sources of funding for their operations, but neither of them conducts extensive archaeometric research.

Other Activities

In addition to conducting research, the SCMRE staff is active in providing exhibits, courses, and workshops for an audience ranging from the bilingual public of California to university students of museum conservation and curators of museums throughout the United States and the world. SCMRE offers internships and fellowships for all educational levels and 15-20 professional courses each year in artifact analysis, preservation, and materials science. It also provides technical information services to conservation and collection professionals outside SI and advises the general public on the preservation of family heirlooms and other privately owned artifacts on request.

SMITHSONIAN ENVIRONMENTAL RESEARCH CENTER

SERC occupies 2700 acres adjoining the Chesapeake Bay in Edgewater, Maryland, about 25 miles from the Smithsonian complex on the Mall in Washington, DC. Employing 14 principal PhD scientists, SERC is one of three Smithsonian research centers (along with SAO and STRI) that function essentially like academic departments or laboratories, with research, educational, and outreach obligations.

SERC scientists perform research in environmental science; estuarine, watershed, and coastal ecology; and climate change, especially its effects on ecosystems. In its words,

> SERC is dedicated to increasing knowledge of the biological and physical processes that sustain life on earth. SERC's interdisciplinary research applies long-term studies to examine the ecological questions about landscapes of linked ecosystems, especially those impacted by human activities.

SERC is the youngest of the Smithsonian organizations, but it has achieved an internationally recognized reputation for outstanding research on environmental problems and the ecology of the land-sea interface since its founding in 1965. The 14 principal investigators are supported by a staff of about 80 (federally supported and trust supported) on the SERC campus in Edgewater.

Research Support and Research Output

Research at SERC is supported by a combination of direct federal appropriations and funding from a variety of other sources, including federal, state, and local grants and contracts received by the principal investigators. In FY 2001, $1.2 million (36%) of the $3.3 million direct federal appropriation was spent on administrative, education, and other infrastructure costs (Table 2-6). The remainder of the direct appropriation was used for research, of which 82% was spent on staff salaries and benefits. The salaries of all but one staff scientist are paid by the federal appropriation. In addition to the federal funds, SERC received about $2.7 million from contracts and grants and $1.2 million from donations and other SI

TABLE 2-6 Estimated Research Expenses of SERC by Source for FY 2001 (in Millions of Dollars)

	Federal Appropriations	Federal—Other	Government Grants and Contracts	Other Trust	Total
Research	2.1	0.1	2.2	0.8	5.2
Salaries and benefits	82%	33%	47%	46%	61%
Other research costs	18%	67%	53%	54%	39%
Other expenses[a]	1.2	0	0.5	0.4	2.1
TOTAL	3.3	0.1	2.7	1.2	7.3

[a]Includes expenses for exhibits and education, administration, and facilities.

sources. The sources of those grants and contracts included NSF, DOC, DOD, DOE, DOI, the Department of Transportation, the National Oceanic and Atmospheric Administration (NOAA), and the Chesapeake Bay Consortium, Inc.

The 14 SERC scientists published 239 articles in 1995-2000. SERC lists articles in 73 journals, including *Science, Nature, Ecology, Limnology and Oceanography, American Naturalist, Geophysical Research Letters, Plant Physiology, BioScience,* and other high-impact journals. SERC authors also wrote or contributed to 36 books in the same period.

SERC, like other Smithsonian units, performs annual personnel evaluations and has fully external peer review evaluations at 3- to 7-year intervals for all promotions. The system in place for staff evaluation entails solicitation of external reviews by mail and their evaluation by a panel of three outside scientists. In addition, SERC has an external visiting review committee that periodically evaluates the relationships between SERC research programs and the broader ecological community and its overall program performance. SERC was last reviewed in 1997.

Unique Characteristics and Special Contributions

One of seven major estuarine research facilities surrounding Chesapeake Bay, SERC is distinguished by its focus on the entire watershed of the bay, with a balanced program of marine, freshwater, and terrestrial studies. SERC scientists conduct research on wetland and forest plants, estuarine microorganisms, ecosystem responses to rising atmospheric CO_2 concentrations and ultraviolet radiation, mangrove forests, primary production by phytoplankton, fish and invertebrate population dynamics, nutrient transport and eutrophication, migratory and wetland birds, carbon storage, and invasive species. SERC scientists perform their field research on the SERC campus, throughout the Chesapeake Bay region, and in 20 states and 23 countries. Also notable is SERC research at the other Smithsonian marine and tropical research centers, the Smithsonian Marine Station at Fort Pierce in Florida and the Carrie-Bow Marine Field Station in Belize. By exploiting the Smithsonian's unique network of marine laboratories in temperate, subtropical, and tropical latitudes, SERC scientists are able to pursue ecological and environmental topics on a wide variety of spatial scales, from local to intercontinental, and in a variety of landscapes and ecosystems. That cosmopolitan scope makes SERC unique among its peer coastal and estuarine institutions around Chesapeake Bay and across the nation.

Another notable aspect of SERC research is an unusual concentration on long-term studies. (See Appendix E for examples.) Once dismissed as monitoring, research carried out over multiple years or even decades is now recognized as crucial to both scientific understanding and the ability to make informed policy about climate change, land management, and maintenance of surface and groundwater quality. The pervasive influence of El Niño on weather around the planet is a familiar example of how disconnected or short-term observations fail to penetrate the dynamics of climate forcing and ecosystem response to El Niño Southern Oscillation variability. Longer-term trends in nutrient delivery to coastal systems and secular trends in atmospheric CO_2 increase and ozone depletion provide further examples of the need for sustained, long-term observations.

Some of the important accomplishments of SERC scientists include:

- Recognition by the journal *Science* as one of the Smithsonian's crown jewels (Pennisi, 2001a) with the highest publication citation impact among SI centers (15 citations per paper in plant and animal science and 13 citations per paper in environment and ecology; Pennisi, 2001b).
- Recognition as the national center for research on biological invasions in marine ecosystems, including establishment of the National Ballast Water Information Clearinghouse, a key information source for the US Congress.
- Invention and patenting of the spectral radiometer, the current national standard for monitoring solar radiation and for systematic research and observations on spectrally resolved ultraviolet radiation.
- Establishment of the ability of riparian zones to remove nitrates from groundwater moving from agricultural fields to streams and estuaries through long-term studies of an entire watershed.
- Conduct of the only long-term experimental CO_2 enrichment of an estuarine marsh to study the effects of rising atmospheric CO_2 on wetlands. This is the world's longest running field experiment on the effects of CO_2 enrichment on entire plant communities.
- Maintenance of extensive geographic databases on the topography, land-use composition and patterns, streams and rivers, and shorelines of the Rhode River watershed.
- Maintenance of long-term monitoring data (10-32 years) on numerous physical, chemical, and biological properties of the Rhode River, a subestuary of the Chesapeake Bay, to test hypotheses about the effects of land use in the watershed on the river, estuary, and bay. Although it is common for marine laboratories like SERC and its counterparts around the bay to maintain local time series on basic marine properties (such as meteorology, salinity, temperature, and chlorophyll), the scope, depth, and range of SERC's emphasis on investigation of long-period, low-frequency processes are unique.

Other Activities

Like its peer institutions and counterparts, SERC engages in a variety of educational and outreach activities. Although SERC is not a degree-granting institution, it hosts undergraduate and graduate students from other institutions who work on the SERC campus and in collaboration with SERC principal investigators. The most important of SERC's educational activities is its fellowship program, which provides support for graduate students and postdoctoral and senior fellows. The SERC program supports 11 fellows. Like that of STRI, SERC's contribution to science education therefore rivals or surpasses fellowship support provided by many educational institutions of comparable size. SERC also has an extensive K-12 education program, which brings 10,000 elementary school students and their teachers to the campus each year for an environmental education program.

3

Findings, Conclusions, and Recommendations

The Committee's analysis of its statement of task, its research into the nature, funding, and work of the six Smithsonian research centers covered by this study, and its consultation with several federal agencies led it to various findings, conclusions, and recommendations, which are presented in this chapter.

It should be noted that there is considerable potential for confusion as to what is meant by the term *research budget*. OMB's table on the allocation of federal research funding (Table 1-1), for example, gives the total FY 2001 federal research appropriation to SI as $108 million. According to SI, however, that figure includes funding for research at SI units other than the six addressed in this study (about $20 million) and overhead costs for all SI research (some $35-40 million). The portion of the federally appropriated research budget allocated to the six science units when added together is about $52.5 million; this amount is divided among the six units as shown in the first columns of Tables 2-1 through 2-6. Also as shown in those tables, the budget of each unit is divided into a "research" component (top line) and an "other expenses" component (second line). (The nature of the "other expenses" is described in the footnotes of the tables.) The budget of each unit is also supplemented with trust funds, and the total budget for all six units from all sources amounts to just over $127 million (Table 1-2). In the following discussions, the term *research budget* refers to the federally appropriated amounts shown on the top line of each table unless otherwise specified.

FINDINGS AND CONCLUSIONS

The six research centers, taken together, embody an important component of SI's research program and constitute a mechanism whereby SI carries out its charter to "increase and diffuse knowledge." The Committee considered the work of each SI unit, its role and status in the scientific enterprise, and whether the terms *uniqueness* and *special contribution* should be applied to each research center. In arriving at its findings and conclusions, the Committee drew on information received from the central offices of the Smithsonian and the research centers themselves, data-gathering interviews with SI staff and representatives of the research centers, the expertise and relevant knowledge of the Committee members themselves, and informal contact with members of the wider scientific community.

A: The research performed by the National Museum of Natural History, the National Zoological Park, and the Smithsonian Center for Materials Research and Education is inextricable from their missions and is appropriately characterized by the terms *unique* and *special contributions*.

The terms of the creation of NMNH make it the nation's repository for extensive collections of gems, minerals, meteorites, plants, animals, fossils, and other natural history specimens. NMNH has a responsibility to conduct collection-based research to derive knowledge and meaning from the collections not only for the sake of scientific advances, but also to enhance the management and public display of the materials. Without collection-based research, NMNH could become simply a warehouse of artifacts and stale information that would quickly fall out of step with advances in the sciences and ultimately fail as a source of public education and attraction. In addition, NMNH has a unique and critical role in the national natural history museum community with collections vastly larger and wider in scope than those of any comparable US institution. The breadth of its research mission and the extent of its service to the museum research community are correspondingly critical.

Similarly, research by NZP is essential to its mission, allowing enhanced management and display of its live animal collections and the development of improved conservation practices for wild animal populations. Research is also an expectation under the terms of the zoo's certification by AZA and other agreements that the zoo has entered into. Although large research programs are not a requirement for AZA accreditation, zoos are expected to have a long-term commitment to conservation and research in proportion to the size and significance of their collections. The Committee, therefore, believes that the permanent loss of the CRC conservation and research budget would not go unnoticed by AZA and could ultimately jeopardize the accreditation status and viability of NZP.

SCMRE occupies a unique niche for highly specialized research that is vital to the work of the SI museums and that does not fit the mission of any other federal or state science agency or of most other privately funded entities. Because its work is paralleled by few institutions, this unit is highly valuable not just to the Smithsonian museums but to the museum community as a whole.

B: The Smithsonian Astrophysical Observatory, the Smithsonian Environmental Research Center, and the Smithsonian Tropical Research Institute are world-class scientific institutions that combine facilities, personnel, and opportunities for specialized long-term research that is enabled by the stability of federal support. These units are engaged in research that supports the mission of the Smithsonian Institution as a whole—increasing knowledge and providing supporting expertise for the activities of other SI units, including educational activities.

The Committee does not believe that the terms *uniqueness* and *special contribution* apply to the work of SAO, STRI, and SERC in the same way that they apply to the research of NMNH, NZP, and SCMRE. The Committee found, however, that SAO, STRI, and SERC are performing scientific research that is of high quality and is well respected by standard measures of scientific output and impact. For example, Pennisi (2001b) reported that STRI, SERC, and SAO rank in the top 1% of institutions in terms of their scientific impact as measured by the number of publications they produce and the number of citations that each paper receives. In addition, SAO, SERC, and STRI all have aspects that make them unique in the nation's research enterprise, allowing research to be carried out on instruments unavailable elsewhere or in unique environments.

SAO, for example, supports path-breaking and high-risk projects, such as the development of gamma-ray astronomy before its general acceptance as an important discipline; the concept of combining multiple mirrors into a single telescope, which led to the MMT; the development and operation of SMA, which will be the only interferometer operating in this wavelength regime until the new Atacama large-millimeter array, now under development, becomes operational; and the initial support of the red-shift survey, which led to the first quantitative mapping of the large-scale structure of the universe and the distribution of galaxies. SAO is active in nearly every field of astronomy and astrophysics; through the availability of its research and facilities, it makes a unique contribution to the Smithsonian's mission and to the furtherance of astronomy in general.

SERC is unique among field stations because of the Smithsonian's ownership of 2700 acres of land, including 12 miles of undeveloped shoreline, and complete control of the Rhode River watershed. Taking advantage of the SERC site, staff scientists fill an important niche in environmental research by emphasizing long-term mechanistic ecosystem studies in a balanced program of watershed, marine, and terrestrial ecology. SERC maintains the largest group of scientists working together in a single location on multidisciplinary investigations of an integrated set of ecosystems subject to most of the environmental and natural-resource management issues faced by the nation today. They study a coastal watershed that is connected to an estuary, and their work is unusual by virtue of the variety of integrated ecosystems and land-use types in the watershed and the access to a protected site with appropriate instrumentation for long-term, landscape-scale studies. The emphases of SERC's investigation of long-period, low-frequency processes are outstanding among research programs in these disciplines.

STRI holds custodianship of Barro Colorado Island under a special agreement with the government of Panama and owns a complex of nine research stations throughout Panama. Those factors allow it to maintain the broadest research program of any US tropical research institution. Staff scientists conduct long-term integrated studies of tropical ecosystems that emphasize the understanding of biodiversity in marine and terrestrial systems and long-term tropical forest dynamics. STRI's work on tropical forests and coral reefs is considered a cornerstone in tropical and marine ecology (Pennisi, 2001b). STRI is also renowned for its role in the training of tropical biologists from around the world.

Although the Committee did not systematically investigate interactions among the science units, the opportunities for such interactions are obvious. For example, the ecologically oriented research of SERC and STRI are clearly complementary to the work of NMNH and NZP in systematics, evolutionary biology, and biodiversity conservation, and SAO and NMNH share work that focuses on the search for extrasolar planets and the evolution of solar systems. Thus, the research of SERC, STRI, and SAO can contribute not only to their own public outreach and education missions but also to the extensive public service provided by the museum and zoo.

C: Funding for research at the Smithsonian's research centers comes from a mix of sources, including a substantial fraction received through open competitive programs.

Direct federal appropriations to the Smithsonian research centers are used primarily as core support for salary and infrastructure costs, with 63-84% going to salaries across the six centers (see Tables 2-1 through 2-6). Doing science, however, involves many other costs, including costs for equipment, supplies, computing, travel (of particular importance to field biologists), and support of graduate students, postdoctoral fellows, and research assistants. At SAO, SERC, STRI, and NZP, appropriated funding is not, for the most part, being used to pay for those other research costs. And although SI can raise private-sector funds and has endowment resources, the degree to which these resources supplement research budgets is small.

The majority of supplemental research funding for the SI science centers comes from competitively awarded grants and contracts. Researchers at SAO, SERC, STRI, and, to some extent, NZP have been highly successful in competing for grants and contracts from other government agencies. In FY 1999 and FY 2000, for example, success rates of grant applications were roughly 56% and 67%, respectively, for SERC, 77% and 47% for STRI, and 50% and 60% for NZP.[1] (Data on success rates could not be compiled for SAO, but 59% of SAO's $112 million comes from government grants and contracts. See Table 2-2.) NMNH's opportunity to compete for research contracts and grants has apparently been limited because NSF, the agency most likely to support research on topics relevant to the work of the museum, has placed restrictions on the funding of federal employees at NMNH. However, NMNH scientists were successful in 67% and 79% of the grants that they applied for in FY 1999 and FY 2000, respectively, with many of the awards coming from private foundations and other nonfederal sources. In contrast, the researchers at SCMRE derived less than $50,000 from competition for government grants in FY 2001, probably because little of its research overlaps with the missions of the granting agencies. In as much as the research centers are already competing for extramural grants to leverage the federal research funds that go mostly to salary support, these research centers are not functionally "exempt" from competitive peer-reviewed grant programs.

The success of many Smithsonian scientists and managers in attracting external funds to their institutions reflects the quality and stature of SI's scientific research and personnel. The research centers indicated that the two factors critical in their search for external funding are the quality and creativity of the proposed research and the association with SI and

[1]These estimates are based on the list of proposals submitted and the list of awards received by SI scientists in FY 1999 and FY 2000 provided by SI. Similar data were not available for SAO.

with SI's worldwide reputation for having the highest standards of scholarship and scientific accomplishment.

Although the Smithsonian is unique in many respects, its research funding structure is analogous to that of many other research entities, such as state-supported universities, in which direct government appropriations provide salary support for the scientific faculty and pay for utilities, space costs, and some fraction of administrative overhead. To do science, additional resources must be secured from other sources. In both public and private universities, in which state-appropriated or endowment funds typically provide some salary and operational costs, external funds from a variety of sources must also be raised to conduct research.

The Committee did not find that Smithsonian researchers' receipt of full (12-month) salary support constitutes a substantial advantage over university-based researchers. The latter usually receive 9-month salaries and obtain funds for another 2 or 3 months from contracts and grants. Some universities even provide 12-month salaries. Given the wide variations in salary among institutions and the need for both SI and academic researchers to compete for resources for computing, travel, and graduate students and postdoctoral researchers, any advantage of SI scientists in this regard would be small. The NAPA study (NAPA 2002) is expected to address those issues in greater detail. There also appears to be little consistent competitive advantage for federally funded scientists at the Smithsonian over federally funded scientists at NASA centers, USDA, or NSF-supported or NASA-supported national observatories, such as the National Optical Astronomy Observatories (NOAO), the National Radio Astronomy Observatory (NRAO), the Space Telescope Science Institute (STScI), and the National Atmospheric and Ionospheric Center (NAIC).

D: The Smithsonian Institution plays an important role in the overall US research enterprise and contributes to the healthy diversity of the nation's scientific enterprise.

Numerous reviews of the nation's scientific enterprise have concluded that the diversity of sources of funding for science and the diversity of institutions that conduct science, including the federal government itself, are good for the overall health of the US scientific enterprise (e.g., National Research Council, 1994). The federal government is the major supporter of basic research in the United States, and its role is crucial in many of the fields in which SI conducts research, such as astronomy, meteoritics, and ecology. In such fields, commercial applications are often remote from the effort to understand basic and general principles, and private funding can be hard to identify. (The Committee recognizes, however, the important role played by foundations and other private sources in the construction and commissioning of ground-based astronomical observatories.)

Prospective merit review is the favored model for allocating federal funding for science, but it is not the only model, nor is it the dominant approach in some sectors of government-supported science, such as defense-related research and development (R&D). Cook-Deegan (1997) characterizes two alternative approaches—"formula-funding" methods, based on political, historical, or performance factors, and what might be called the "DARPA" or "strong-manager" model, in which staff experts decide how to distribute research funds. The strong-manager model is most developed in the realm of the DOD, and both the Defense Advanced Research Projects Agency (DARPA) and the Office of Naval Research, where this model flourishes, have achieved spectacular successes (particularly in information and communication technologies) using this approach when "outside-the-box thinking" is required to develop innovations.

Although this year's guidance to the science-funding agencies on R&D priorities from John H. Marburger, the current director of OSTP, and Mitchell Daniels, the current director of OMB, reaffirms the preference for merit-based competitive peer review, it also contains provisions for exceptions relying on other models.

> Programs should maximize the quality of R&D they fund through the use of a clearly stated, defensible method for awarding a significant majority of their funding. A customary method for promoting R&D quality is the use of a competitive, merit-based process. NSF's process for peer-reviewed, competitive award of its R&D grants is a good example. Justifications for processes other than competitive merit review may include "outside the box" thinking, a need for timelineness, *unique skills or facilities, or a proven record of outstanding performance* [emphasis added; Marburger and Daniels, 2002].

The Committee believes that the Smithsonian science units exhibit "unique skills and facilities" and "a proven record of outstanding performance" and thus meet the criteria set forth in the FY 2004 budget guidance for use of a process other than peer review to allocate its research resources. Moreover, the Committee notes that a diversity of funding models and mechanisms can help to ensure that resources (such as properties belonging to or under the management of SERC and STRI or some of the instrumentation available at SAO) that may not lend themselves readily to use under the standard model for prospective merit review because of limitations of ownership or legal agreements are nevertheless used, and used effectively, to contribute to the nation's scientific enterprise.

It is, of course, fair to ask that research carried out with uncompeted federal support meet the same standards of excellence via retrospective merit review that are applied to research performed in university laboratories and industrial facilities. Both public and private research institutions should be evaluated according to similar standards and judged by quality and relevance, not only according to whether their research has been subjected to prospective peer review.

E: Mechanisms at the Smithsonian scientific research centers for evaluating overall scientific productivity and for evaluating the productivity of individual scientists are variable and inconsistent.

SI's use of visiting committees made up of experts in the relevant research for programmatic review is similar to review of in-house or intramural research at other federal agencies. Past external reviews of the six SI research centers have assessed the quality of the research programs and their relevance to the institution's mission. Those external reviews are very similar to the "expert review" that has been recognized by the greater scientific community as the most effective means of evaluating federally funded research programs (National Academy of Sciences/National Academy of Engineering/Institute of Medicine, 1999, 2001). The use of national searches for recruiting new scientists and the solicitation of external peer evaluation for individual performance, promotions, and so forth on approximately the same time scale as is common in universities appear to conform with what is expected at research institutions in the United States.

However, the Committee found that the institutional review processes in place were at times lacking in depth (especially at the larger research centers, where reviews of individual departments have not been held regularly and have not involved external input until recently) and in frequency. For example, one external committee of eight members meeting for 2-3 days can hardly conduct a comprehensive review of the vast array of programs at SAO. The Committee noted that SAO is planning to institute a new system of more in-depth divisional review, and it strongly encourages SAO to follow through with its plan. The Committee also considers the newly implemented institutional review system at NMNH to be a big step forward toward comprehensive and insightful review.

F: Communication between the research centers and the central management of the Smithsonian Institution appears to be weak.

The Committee found that communication between the research centers and the central Smithsonian administrative offices in "the Castle" was not as good as it could and should be. The apparent lack of good communication contributes, on the one hand, to some misunderstanding by the government and the general public about the centrality of research to the Smithsonian's mission and, on the other, to the Castle's inability to articulate a strong message about Smithsonian research that matches the high quality of the research programs themselves.

Consequences of Transferring Federally Appropriated Research Funds from the Smithsonian

The remainder of the Committee's findings and conclusions focus on the results of its consideration of the consequences of transferring the federal research funds now appropriated to the Smithsonian to competitive peer-reviewed programs. OMB's original proposal for the FY 2003 budget was that this transfer be made to NSF. Although the Committee recognizes that there may be other agencies conducting peer-reviewed programs that might be appropriate to receive some of SI's work, it has used NSF as its main point of reference, and consideration of potential transfers to other agency destinations is beyond the scope of this study. In addition, given the abbreviated timeframe for completion of its study, the Committee was not able to conduct its analysis at the level of individual programs at the science centers but rather considered only the case of transfer of the research budget of an entire research center to NSF. Clearly, the effects of reallocating SI research funds to NSF would differ according to the terms established for the transfer of funding. For example, the consequences of directing all funding classified by SI as "research" to be transferred to NSF would be much more severe than the consequences of transferring only the portion of the SI's research funds that is not allocated to the support of salary and infrastructure costs. (In the latter case, the amount to be transferred would, in fact, be modest.) Of course, transfer of the entire federal appropriation—the amounts listed as "Totals" in Tables 2-1 through 2-6—would be extreme. Similarly, if the funds were transferred to NSF without any directives for their use, the stability of the SI programs would be much more threatened than if the transfers were made with directives to NSF to maintain the programs and simply open them up to competition. No details of suggested terms for the transfer of funds were provided to the Committee, and the time allotted for this study did not allow for a comprehensive consideration of all possible scenarios. The Committee elected to focus on four reasonable scenarios in evaluating implications: (1) transfer to NSF of all federally appropriated research funds, (2) transfer to NSF of only funds not allocated to salaries, (3) transfer to NSF of funds with no requirement to continue to invest the funds in the same or similar programs, and (4) transfer to NSF of funds with directives to maintain the programs but allow competition from non-Smithsonian persons and entities.

As stated in Chapter 1, the Committee was not charged to assess the quality of the research at each of the six centers *per se* or whether the research programs fit SI's mission. Those issues are currently being evaluated by the Smithsonian Science Commission, which is expected to deliver its report to the Board of Regents at the end of 2002.

G: **In general, transfer of all federal research funds (including salary and, in some cases, infrastructure support) would greatly reduce and possibly eliminate the role of the federal government in the long-term support of the core scientific research staff who provide the foundation of the Smithsonian research program. A withdrawal of federal support of this magnitude would make maintaining the staff and programs of the centers extremely difficult and would very likely lead to the demise of much of the Smithsonian's scientific research program.**

In the event of the transfer of all federal research funds, including salary and infrastructure support, from SI, the funds would have to be replaced by trust funds (i.e. funds other than direct federal appropriations and transfers) if the SI science centers were to continue to operate as usual. The fraction of research budgets now provided by trust funds varies from virtually zero for SCMRE (which received only about $50,000 in trust funds in FY 2001) to about 70% for SAO. Given the current decline in SI income from commercial activities (gift shop and magazine sales) and other sources, it is likely that to survive, the research centers would have to replace the lost federal funds with outside funding from contracts and grants. The fact that small investigator grants typically cover no more than 2 or 3 months of salary in many of the scientific fields covered by the six research centers could present a major difficulty even if NSF were to maintain the scientific programs transferred from SI and simply manage them as its own. [Some models among NSF's programs would allow for payment of full salaries. An example is the cooperative agreement under which the National Center for Atmospheric Research (NCAR) is operated by the University Corporation for Atmospheric Research (UCAR). For such a model to reduce effects on SI personnel, SI would have to be able to compete for and win such an agreement.] The results are likely to be, at best, major instability for SI scientific personnel and widespread disruptions to and loss of continuity in on-going programs. The only major exception might be the programs supported by the large NASA contracts at SAO, but even SAO would suffer major effects and experience a profound reduction in the breadth of its scientific programs.

H: **Transferring the federally appropriated research funds for the National Museum of Natural History and the National Zoological Park to competitive programs at the National Science Foundation is likely to jeopardize their standing in the museum and zoo communities and could seriously damage aspects of their nonresearch roles. If the fund transfer were large and included salary support, the positions of critical museum and zoo personnel could be threatened. Loss of core funds could also lead to the closure of the Smithsonian Center for Materials Research and Education.**

Research is an integral part of the missions of NMNH and NZP and is inextricably linked to their effective management of their collections (see Finding A). Furthermore, because the salaries of many personnel critical to the mission of the museum and zoo (such as curators and veterinarians) are paid from the "research" part of their federal appropriations, transfer of the entire federally appropriated "research" budget could translate into severe cuts in the amount of support for the salaries of critical staff even though research is but one component of the roles they play at the museum and the zoo. (NMNH has undergone a reduction of 25% in the number of its curators by attrition over the last 10 years. The academic and research implications for reductions in systematics and taxonomy in the nation, as well as in the Smithsonian, deserve examination by further study groups.) Replacing funds for salaries of mission-critical personnel from the "nonresearch" portion of the museum's and zoo's budget would also probably produce severe funding shortages for their other nonresearch functions and activities.

Transferring only the small portion of appropriated research funding not now allocated to salary and infrastructure at NMNH and NZP would result in smaller effects but is still likely to produce a lower level of investment in the topics addressed by NMNH and NZP, few of which are currently major components of NSF's portfolio. Some of the effects could be offset by directing NSF to maintain programs in the affected fields, but it is not clear that this would be a viable long-term approach. Under such circumstances, it would become particularly important that the eligibility of NMNH personnel to compete for NSF funding be confirmed. In essence, such a smaller-scale transfer could still constitute a removal of the research functions of NMNH and NZP, which could endanger their standing in the museum and zoo communities. It should be noted that all modern zoos and natural history museums—such as the Bronx Zoo, the San Diego Zoo, the American Museum of Natural History, and the California

Academy of Sciences—have their own research facilities. In addition, the transfer of research funding would result in a loss of the vital direct connection among research activities, collection needs, and the missions of the units.

SCMRE was created to perform research. Without its federally appropriated research budget, its ability to fulfill the purpose of its creation would be severely damaged, and it would most likely cease to exist, given that its appropriation represents nearly 90% of its budget. Because SCMRE's research is highly specialized and does not fit the mission of any of the other federal science agencies, it is very unlikely that the lost federal funding could be readily replaced via competitive grants. (Although it is also possible that NSF could be directed to maintain an SCMRE-style program with the transferred funds, this would be a clear mismatch for NSF, which has little in-house expertise in managing such a program.) It is also unlikely that private foundations and granting agencies would be able to make a long-term commitment to sustain SCMRE. SCMRE might be able to continue some of its operations by charging client museums for its services, but this mode of operation would be precarious, at best. The termination of SCMRE could have serious repercussions for the Smithsonian's museums and the museum community as a whole. The hundreds of material analyses that SCMRE performs for the Smithsonian museums each year would have to be either contracted out or performed by museum staff; this would burden museum budgets, and it might not even be feasible if museum staff lack the technical expertise necessary for the more sophisticated preservation and conservation work. In addition, SCMRE helps to place museum objects in contextual frameworks that the public can relate to, thereby making exhibits more lively and interesting. The benefits that SCMRE provides to both its professional community and the public would probably be lost.

In short, the Committee believes that transferring the research funding of NMNH, NZP, and SCMRE to NSF would have numerous harmful effects. Given that it is the Committee's judgment that the research performed by NMNH, NZP, and SCMRE is an intrinsic part of their missions and is justifiably characterized by the terms *unique* and *special contribution* (Finding A), the Committee believes that such a transfer would be inappropriate and that these centers should continue to be exempt from open competition for research funding.

I: Transferring directly appropriated funds from the Smithsonian Astrophysical Observatory, the Smithsonian Environmental Research Center, and the Smithsonian Tropical Research Institute to a competitive mechanism while trying to maintain the centers in the

Smithsonian could produce consequences ranging from moderately or seriously deleterious to termination of their operations.

In evaluating the consequences of a transfer of federally appropriated research funds from SAO, SERC, and STRI to NSF, the Committee considered two cases: (1) transfer of all federally appropriated research funds and (2) transfer of only the funds not allocated to salaries. Although the federal research appropriation makes up only a portion of the total research budgets for the three science centers, it provides a strong foundation for their ability to participate broadly in the research enterprise, with the majority of the appropriated funding going toward staff salaries (63% in SAO, 74% in STRI, and 82% in SERC; see Tables 2-2, 2-4, and 2-6, respectively). The supported staff serve as "nucleation sites" for larger research efforts. The federally appropriated funding is highly leveraged, and there is extremely little discretionary money from the appropriated funds per research staff member.

In the first case, the loss of all federally appropriated research funding would severely erode the ability of the centers to support their core staff. Although all three of the centers bring in substantial funding through competitive grants and contracts, increasing this category of support is not necessarily a reliable means of covering the lost funding for full salaries, as noted earlier. Trust funds other than federally funded contracts and grants would have to make up the difference, or staff would have to be let go. Under such circumstances, the centers might cease to function as world-renowned centers of excellence.

Although SAO is far less dependent on the direct federal appropriation for salaries because of its relatively large supplemental income from NASA contracts, the federal funding does support about 64 SAO full-time-equivalent staff scientists who engage in a large variety of astronomy and astrophysics work that far exceeds the scope of what is performed for NASA. SAO's breadth of expertise has, in fact, been the foundation on which SAO operations were built. For instance, the long-term expertise in the high-energy group was fundamental to the development of the Chandra project, which led to the contract awarded by NASA. Although SAO might survive on only its contract and grant income, it is likely that its current role as the multifaceted long-term research center it is today would be difficult to maintain with only outside funding.

Marine and estuarine research is reasonably, albeit not lavishly, funded in the overall US scientific community. There is a tradition in the marine sciences of supporting quasi-soft-money research institutions, such as the Scripps Institution of Oceanography, the Woods Hole Oceanographic Institution, and parts of the Marine Biological Laboratory (MBL). These organizations typically have large open-ocean research components with substantial funds allocated to maintain research vessels and physical oceanographic research. Each also has existing relationships with larger academic or government institutions that provide support for infrastructure, salary support, appointments for faculty, and so on. It is, however, extremely rare for such institutions to focus on estuarine and coastal issues, as does SERC. In addition, it is not clear whether there is sufficient funding overall in this community to absorb one more soft-money institution. It might be possible for SERC to maintain its research continuity and productivity in the marine sciences community, but it is likely to be difficult.

Unlike the situation in the marine sciences, few research institutions in terrestrial ecology and environmental science are supported solely through extramural funds. There are some examples, such as the Ecosystems Research Center (part of MBL), the Woods Hole Research Center (WHRC), and the Natural Resources Ecology Laboratory (a research center within Colorado State University). But with the exception of WHRC, each has an existing relationship or is part of a larger institution that provides support and services. Each is also substantially smaller than STRI in numbers of senior scientists and total staff. WHRC and MBL have endowments that have been built up over periods of years, and all are active fundraisers. STRI has also been a successful fundraiser, but its opportunity to build an endowment has been limited, and this limitation could pose a substantial hurdle. Perhaps most important, research budgets for terrestrial ecology, environmental science, and conservation biology are not as large as for the marine sciences and have not been growing. There is little room for adding independent institutions with needs for administrative and operational support as well as support for the actual science to the mix.

The impacts of the second case, transferring only the parts of the federally appropriated research budgets that are not allocated to salaries, could be much less significant. The scientific staffs of SERC, STRI, and SAO have impressive track records in winning grants and contracts, and presumably could continue to do so for other research expenses. They might, however, have less discretion in their choices of research topics, given that NSF might decide not to entertain proposals dealing with some of the long-term work in which they are now engaged. NSF could, of course, be directed to maintain the SI programs and merely open them to competition, which might ensure that some of SI's research programs would continue.

The least damaging alternatives for all three science units could involve transfer of their federal research funds to NSF with the understanding that NSF would continue to operate the research units under contracts or cooperative agreements awarded via peer-reviewed competitions. Such contracts or agreements can include provisions that allow the funds that NSF awards to be used to pay salaries, infrastructure costs, and so on. For example, NSF's cooperative agreement with UCAR provides funds for NCAR staff salaries, operation and maintenance of equipment and campus buildings, and other fundamental expenses. For such terms to hold out at least the possibility of reducing impacts on SERC, STRI, and SAO to a minimum, it would be necessary to allow the Smithsonian to engage in open competition. There could, of course, be many practical considerations that could create difficulties that the Committee is not equipped to analyze in any depth. Would, for example, the ownership of the SERC properties by the Smithsonian pose serious barriers to NSF management? (Typically, the facilities operated by NSF in this way belong to NSF.) Would the agreements between the Panamanian government and the Smithsonian preclude or hamper transferring the management of STRI to NSF? And how likely is it that the uncertainties associated with the need for SI to compete to manage and operate its facilities under contracts or agreements with NSF would drive key staff away before the outcome of the competition was known? These are only some of the many issues that would need to be considered if the proposal to transfer the research funding of SAO, STRI, and SERC to NSF were pursued.

J: The Committee could not identify any substantial advantages with respect to organization, management, or quality assurance that would accrue from changing the current system of federally appropriated research funding for the Smithsonian Astrophysical Observatory, the Smithsonian Environmental Research Center, and the Smithsonian Tropical Research Institute.

If SAO, SERC, and STRI were maintained more or less intact under the direct management of NSF or via an agency-sponsored consortium and the scientists that they house continued to seek outside research support as they do now, it is difficult to see that there would be much net cost or benefit to the overall national scientific enterprise. However, there would still be organizational risks. NSF, in particular, generally operates research facilities that are clearly called for in a national process that has reached out to the relevant research communities. Typically, NSF-operated facilities are ones that would be difficult or uneconomical for single universities to sustain by themselves—for example, large space-based or ground-based telescopes, research vessels, aircraft, and state-of-the-art supercomputers. NSF usually supports such facilities and their research infrastructures as community resources, so there are explicit rules and opportunities for scientists from all over the country (and in many cases visitors from other countries) to use them. STRI has those features to some degree, but SAO and SERC do not, although they have active and productive programs for enabling visiting scientists to maintain research projects in their nearby field sites. Because it would be difficult to justify the Smithsonian astrophysical, environmental, and tropical research laboratories as national facilities, in the way that the NCAR or the Antarctic research stations are justified, there is a risk that they would be perceived as out of place in the overall mission of NSF.

For example, the Committee considered SAO and the nature of the national observatories funded by NSF. The major national astronomy user facilities use a variety of management models. NOAO is managed by Associated Universities for Research in Astronomy (AURA), a not-for-profit corporation, under a cooperative agreement with NSF. NRAO is managed by Associated Universities, Incorporated, also under a cooperative agreement with NSF. The Hubble Space Telescope is operated by the STScI, also set up by AURA under a contract with NASA. NAIC at Arecibo is managed by Cornell University with substantial external oversight. These models have various levels of community oversight, usually including community involvement in their governance (for example, on their boards of directors or oversight councils, visiting committees, search committees for observatory directors, user committees, and peer-review committees) and concurrence of a contract-mandated sponsoring agency (NSF or NASA) on major policy and governance issues.[2] Because SAO is not a service organization, none of those models would be truly appropriate for SAO without a profound restructuring of its mission and direction. Only a modest fraction of SAO's current activities could be restructured into unique national "user facilities" that could justify support from NSF. The Committee sees no benefit of such an arrangement.

The Committee could see no management or organizational advantage, or any question as to the current quality of the science that these centers are producing, that would warrant changing the current system of federally appropriated research funding in support of SAO, SERC, and STRI through the Smithsonian. Rather, the Committee believes that the benefits of opening up their research programs to competition would be so marginal as to be outweighed by the costs in uncertainty and disruption. The Committee is not claiming that the Smithsonian's management of these science units is without flaws; for example, shortcomings in communication and reporting within and between the centers and SI were obvious to the Committee. But the Smithsonian and its research centers are addressing these issues with the assistance of the Smithsonian Science Commission at the institution level and in a series of efforts at the center

[2]Note that the Chandra X-Ray Observatory is operated by the Chandra Science Center at SAO itself, under contract with NASA; it has users and peer-review committees established in accordance with its NASA contract.

level, such as SAO's Long-Range Planning Committee and Issues Committee. Such efforts should be allowed to come to completion, and the changes recommended should be implemented and tested.

K: The Committee identified little or no scientific benefit of transferring federal funds away from the Smithsonian. The implications for the relevant scientific fields are likely to be adverse.

The disruption of scientists and their activities at the six research centers would cost the scientific community some progress and lead to some setback of their fields of research. For example, a substantial potential danger of switching SAO, SERC, and STRI into the soft money arena is the disruption or end of valuable long-term research that has been judged by external review to be excellent. The core support of the SAO x-ray astronomy group in the 1970s and 1980s, for instance, led not only to productive research with predecessor facilities but also to the development of the concepts and technology that enabled the building and operation of the Chandra X-Ray Observatory, the foremost x-ray telescope in the world and a productive national facility available to all astronomers. That experience has been repeated in other fields of astronomy at SAO.

Similarly, much of SERC's and STRI's research emphasizes studies that depend on long-term monitoring, a category of projects that has never fared well in open competition for 3-year grants but whose value is becoming ever clearer in light of global environmental change. In a scenario in which SI would be left with sufficient resources to support the salaries of the SERC and STRI staff, these long-term programs could plausibly continue to be funded by NSF, although none of its current programs would accommodate them without substantial changes in policy. For example, NSF's Long Term Research in Environmental Biology (LTREB) program is designed to provide funding to help maintain continuing long-term research projects. Awards for this program are, however, made only to hypothesis-driven projects and are limited to a maximum of $60,000 per year for up to 5 years (maximal total award, $300,000). The LTREB program does not support basic monitoring efforts. A more appropriate option might be NSF's 6-year renewable Long Term Ecological Research (LTER) program, which emphasizes long-term studies. Under its current policies, however, the LTER program does not consider unsolicited proposals and accepts proposals only during periodic open competitions. The last such competition for LTER funds was held in 2000, and no date has been set for another.[3] Unless NSF were directed to continue support for

---

[3]The Committee consulted the program director of LTER regarding the format and date of an LTER competition.

SI's long-term environmental projects as a priority component of the LTER or another appropriate program, a transfer of funds to NSF could result in a loss of funding for some or all of SI's valuable long-term studies. Given the vast array of long-term monitoring undertaken by SERC and STRI, some projects might not be suitable for LTER competition. However, it should be noted that NSF's advisory body, the National Science Board, recently recommended that NSF substantially increase its investments in existing long-term research programs and establish new support mechanisms for additional long-term research (National Science Board, 2000). Of course, even if NSF were to commit to continued funding for SI's long-term studies, SI scientists would have to compete for the right to continue to work on them, assuming that no restrictions were placed on their doing so.

Would the transfer of funds to open competition without a commitment to maintain the existing facilities benefit the field and other researchers? In the Committee's judgment, it is quite unlikely to do so. There would no doubt be some benefit in having some "extra" funds available in appropriate competitions and panels, and some of the researchers now at the Smithsonian would no doubt continue to be successful in competing for them. But science is not accomplished only by winning competitive grants. It also requires infrastructure, planning, facilities, and a reasonably stable administrative structure for which the support of productive scientists has high priority. These more operational and programmatic needs would certainly be endangered, with slim prospects for replacing them or making up for them. In this case, the overall effect on the relevant scientific fields is likely to be adverse. Even with a commitment to maintain the facilities under NSF management, disruption, uncertainty, and the need for SI to divert resources to engage in the competitive process are likely to reduce productivity and erode morale for benefits that again, in the Committee's view, are far outweighed by costs.

L: The broad mission of the Smithsonian Institution would be compromised if the links between the Smithsonian and its research centers were broken by transferring sponsorship of the centers to the National Science Foundation.

The work of the six scientific research centers is compatible with and legitimized by the Smithsonian's charter to "increase and diffuse knowledge." Each of the centers relies on its federal support to function effectively. The federal appropriations provide reliable, long-term support for core research staff, just as universities provide support to their tenured professors. If a transfer of sponsorship of the research centers could be carried out in an institutional framework that preserved tenure-like support for the core research staff, the scientific productivity of the research centers might be preserved, but the mission of SI would still be damaged. Severing the link could have serious consequences for the overall reputation of the institution. It is broadly recognized that the association of the six centers with SI helps to leverage support from donors and acts as a magnet for the best professional and administrative staff. Having lost some of its ability to generate knowledge, SI would not be in a good position to recapture it even if its fiscal situation improved. There would certainly be a strong adverse reaction in the scientific community.

Severing the link between SI and its research centers could also have substantial adverse effects on the SI collections, the quality of its exhibits, its international programs and collaborations, and its public-education activities. Entities that have active research programs find it easier to keep in step with advances in the sciences and ultimately to maintain vibrant mechanisms for public education and to remain public attractions. The Smithsonian collections at the museum and zoo could suffer greatly if separated from the research programs through which they maintain currency and relevance. Science is a dynamic, creative process, and collections and exhibits that do not have a chance to benefit from interaction with active scientists have difficulty in maintaining the excitement and interest needed to make them successful. If the Smithsonian did not have its science centers, it would have to invent new links to the scientific community to ensure that the quality of its exhibits and public education did not decline.

RECOMMENDATIONS

1. Research is an intrinsic part of the mission of the National Museum of Natural History and the National Zoological Park. These centers should continue to be exempt from open competition for research funding because of the uniqueness and special contributions conferred by association with their collections.

2. The Smithsonian Center for Materials Research and Education occupies a highly specialized research niche that is of unique and major value to museums of the Smithsonian Institution and to the museum community at large. Hence, the Committee believes that the center should continue to be exempt from open competition for research funding because of its uniqueness and special contributions to the museum community.

3. The Committee believes that the Smithsonian Astrophysical Observatory, the Smithsonian Tropical Research Institute, and the Smithsonian Environmental Research Center should continue to receive federally appropriated research funding. Use of public funds by these facilities is already producing science of the highest quality. Much of the

"research funding" (for other than salary and infrastructure costs) is already obtained via competition. Any benefits of shifting these three facilities to the jurisdiction of another organization would be greatly outweighed by the harm done to their contributions to the relevant scientific fields.

4. Regular in-depth reviews by external advisory committees are essential for maintaining the health, vitality, and scientific excellence of the Smithsonian Institution. Although details of the nature and processes of the reviews may vary to accommodate differences among the six centers, such institutional reviews should be uniformly required for all six Smithsonian science centers and for their individual departments, if warranted by their size. Retrospective external peer review is especially important for areas not routinely engaging in competition for grants and contracts. Regular cycles of review followed by strategic planning offer the best means of ensuring that the quality of SI's science is maintained.

5. The research programs at the Smithsonian Institution provide essential support to the museums and collections, make substantial contributions to the relevant scientific fields, and fulfill the broader Smithsonian mission to "increase and diffuse knowledge." The Committee urges a stronger sense of institutional stewardship for these research programs as integral components of the Smithsonian. The Secretary and the Board of Regents should improve communication with the research centers and become strong advocates for their goals and achievements in a manner that is compelling to the Executive Branch, Congress, and the public.

References

American Zoo and Aquarium Association. 2002. Guide to Accreditation of Zoological Parks and Aquariums.

Cook-Deegan, R. M. 1997, Winter 1996-97. Does NIH need a DARPA? Issues in Science and Technology XIII:25-28.

Gibbons, J. H. and L. Panetta. 1994. Memorandum to Heads of Executive Departments and Agencies. May 6, 1994.

Marburger, J. H. and M. Daniels. 2002. Memorandum to Heads of Executive Departments and Agencies: FY 2004 Integrating Research and Development Priorities. May 30, 2002.

Miller, S. E. 2001. A Smithsonian jewel: Biological collections. Science 293:1433.

National Academy of Public Administration. 2002. Scientific Research at the Smithsonian Institution. Washington, DC: National Academy of Public Administration.

National Academy of Sciences/National Academy of Engineering/Institute of Medicine. 1999. Evaluating Federal Research Programs. Washington, DC: National Academy Press.

National Academy of Sciences/National Academy of Engineering/Institute of Medicine. 2001. Implementing the Government Performance and Results Act for Research. Washington, DC: National Academy Press.

National Research Council. 1994. Investing in the National Research Initiative. Washington, DC: National Academy Press.

National Research Council. 1995. Allocating Federal Funds for Science and Technology. Washington, DC: National Academy Press.

National Research Council. 2000. National Research Initiative: A Vital Competitive Grants Program in Food, Fiber, and Natural-Resources Research. Washington, DC: National Academy Press.

National Science Board. 2000. Environmental Science and Engineering for the 21st Century: The Role of the National Science Foundation. Washington, DC: National Science Foundation.

National Science Foundation. 1980. NSF Circular No. 108 (Revision No. 4). Washington, DC.

Office of Management and Budget. 2002. OMB Circular No. A-11. Washington, DC.

Office of Management and Budget. 2003. Budget of the United States Government, Fiscal Year 2003. Washington, DC: US Government Printing Office.

Pennisi, E. 2001a. Jewels in the crown II: Environment Center. Science 293:197.

Pennisi, E. 2001b. Smithsonian Institution: Turmoil behind the exhibits. Science 293:194-198.

Appendix A

Committee on Smithsonian Scientific Research: Biographical Sketches

Cornelius J. Pings is president emeritus of the Association of American Universities (AAU). He served as president of AAU from 1993 to 1998 and as provost of the University of Southern California from 1981 to 1993. A member of the National Academy of Engineering, Dr. Pings previously served as professor of chemical engineering and chemical physics, vice provost, and dean of graduate studies at California Institute of Technology. He has been elected as a fellow of the American Institute of Chemical Engineering and of the American Academy of Arts and Sciences and is the recipient of numerous awards from organizations, including the American Society for Engineering Education. He has served on several National Academies panels; he was chair of the Committee on Science, Engineering, and Public Policy, cochair of the Fourth Decade Committee, and chair of the Task Group on Institutional Arrangements for Facilitating Research on the International Space Station.

Barbara L. Bedford is a senior research associate at Cornell University. She received her BA from Marquette University in theology, where she was elected to Alpha Sigma Nu, the National Jesuit Honor Society, and her MS and PhD in environmental science from the University of Wisconsin-Madison. For 10 years (1980-1990) she was associate director and for a year (1991) director of the Ecosystems Research Center for Excellence at Cornell University. Before assuming her academic positions, she worked with local and state government agencies in wetlands mapping inventory and classification and development of wetlands regulations. She has been recognized twice by Cornell University for excellence in teaching and in

2001 received the National Merit Award of the Society of Wetland Scientists for outstanding achievements in wetland science. She has served on numerous national committees, including the Management Advisory Group to the Environmental Protection Agency (EPA) Assistant Administrator for Water, the Wetland Experts Team of the Nature Conservancy, and the Technical Oversight Committee for restoration of the Hole-in-the-Donut in Everglades National Park, for which she served as chair. In 1993-1995, she was a member of the National Research Council Committee on Wetland Characterization, and she has served as a consultant on wetlands to EPA's Science Advisory Board. Her current research focuses on plant ecology of freshwater wetlands, especially the biogeochemical and hydrologic controls of plant species diversity on local and regional scales.

Marc Davis is professor of astronomy and physics at the University of California, Berkeley, where he served as chair of the Astronomy Department in 1988-1992. He received his BS from the Massachusetts Institute of Technology and his MA and PhD from Princeton University. His research interests include physical cosmology and large-scale velocity fields. He and members of his research group are working on a DEEP shift survey of the distant universe with the Keck telescope and on generation of maps of galactic dust for use in estimation of reddening and cosmic microwave background radiation foregrounds. Dr. Davis has served on several National Academy of Sciences and National Research Council committees including serving two terms as chair of the Committee on Astronomy and Astrophysics. He has also served as a member of the Visiting Committee for the Harvard Smithsonian Center for Astrophysics.

Hugh W. Ducklow is Glucksman Professor of Marine Science at the College of William and Mary. He received his AB from Harvard College and his AM and PhD from Harvard University. He is studying biological oceanography on marine microbial plankton in habitats ranging from the York River through Chesapeake Bay to the open sea, inland seas, and Antarctic coastal seas. His research focuses on temporal and spatial variations of bacterial biomass, growth dynamics, and organic-matter use. He is active in the Joint Global Ocean Flux Study, where he has been investigating water-column processes in the Chesapeake Bay since 1981.

Jonathan Fink is professor and vice president for research and economic affairs at the Arizona State University. He received his BA from Colby College and his PhD from Stanford University. He has served as director of the Petrology and Geochemistry Program at the National Science Foundation and as chair of a National Research Council committee evaluating the future of the USGS Volcano Hazards Program. His research is concerned with how that magma intrudes into, erupts through, and flows across the earth's surface. His other research interests outside volcanology involve the application of the principles of mechanics to selected problems in tectonics, sedimentology, and planetary geology.

Anthony Janetos is a senior research fellow at the H. John Heinz III Center for Science, Economics and the Environment. Before joining the center in June 2002, he served as senior vice president and chief of program, at the World Resources Institute from 1999. He had served as senior scientist for the Land-Cover and Land-Use Change Program in the National Aeronautics and Space Administration Office of Earth Science and was program scientist for the Landsat 7 mission. Dr. Janetos graduated magna cum laude from Harvard College with a bachelor's degree in biology and earned a master's degree and a PhD in biology from Princeton University. He had many years of experience in managing scientific research programs on a variety of ecological and environmental topics, including air-pollution effects on forests, climate-change effects, land-use change, ecosystem modeling, and the global carbon cycle. He was a cochair of the US National Assessment of the Potential Consequences of Climate Variability and Change and an author of the Global Biodiversity Report and of the IPCC Special Report on Land-Use Change and Forestry.

Kenneth I. Kellermann is chief scientist at the National Radio Astronomy Observatory (NRAO) and a research professor at the University of Virginia. He received an SB in physics from the Massachusetts Institute of Technology and a PhD in Physics and Astronomy from California Institute of Technology. He has served as the assistant director at NRAO and director at the Max Planck Institute for Radio Astronomy in Bonn, Germany. He is a former chair of the National Academy of Sciences (NAS) Astronomy Section, the US National Committee for the International Astronomical Union, and the Radio Astronomy Panel of the Astronomy and Astrophysics Survey Committee and has served on other National Research Council committees, boards, and commissions and on the NAS Council. His research interests include radio galaxies, quasars and cosmology, and the development of new instrumentation for radio astronomy.

J. Patrick Kociolek is curator and G. Dallas Hanna Chair in Diatom Studies at the California Academy of Sciences. He served as the director of research in 1993-1997 and has served as executive director since 1998. He received his BS from St. Mary's College of Maryland, his MS from Bowling Green State University, and his PhD from the University of Michigan.

His research focuses on the taxonomy, ultrastructure, systematics, and phylogeny of the diatoms and on how historical events have determined their distributions over space and time. He is describing the diversity and morphology of the large genus *Gomphonema* (1000+ taxa) and the biogeography and evolutionary relationships of *Actinella* taxa and their allies in the rhaphidioid lineage. As part of his curatorial duties, he is involved in the development of information-management systems that organize and disseminate information on diatom biogeography, nomenclature, and literature. He is also interested in applying results of his studies on diatoms to broader questions of pattern and process in evolutionary biology.

Daniel Livingstone is James B. Duke Professor of Biology and Earth and Ocean Science at Duke University. He received his BSc and MSc from Dalhousie University, Canada, and his PhD from Yale University. He works at the interface of zoology, botany, and geology. He has published papers on chemical embryology, paleontology, fish zoogeography, kinetics of phosphorus cycling, orientation of thaw lakes, management of trout populations, paleolimnology, theory of ice ages, chemical composition of lakes and rivers, folklore, crocodile behavior, geochemical cycles, interactions of climate and human culture, coring technology, and pollen analysis, especially of Alaska, Nova Scotia, and tropical Africa. He is the recipient of the 1989 G. E. Hutchinson Medal of the American Society of Limnology and Oceanography.

Michael J. Novacek is senior vice president, provost of science, and curator at the American Museum of Natural History. He received his PhD in paleontology from the University of California, Berkeley. His research focuses on the higher-level phylogeny of the mammals with emphasis on the radiation of the modern groups of mammals, the placentals, including primates and a great diversity of other major groups. Developing theories largely from paleontological and morphological databases, he has reviewed, analyzed, or incorporated new data from gene sequences. He is part of collaborative efforts to summarize a wide variety of morphological and molecular data to develop a better map of mammal evolution.

Bruce A. Rideout serves the Zoological Society of San Diego as head of the Pathology Division at the Center for Reproduction of Endangered Species. He joined the society in 1991 as an associate pathologist and was named head of the division in 1996. He is also a charter member of the Endangered Species Recovery Council and research associate of the Peregrine Fund. He received an undergraduate degree in chemistry from the University of California, San Diego, and his doctorate in veterinary medicine degree in 1986 from the University of California, Davis. After completing pathology residency training at the National Zoological Park in Washington, DC, he returned to the University of California, Davis, where he received a PhD studying the effects of retroviruses on the immune system. He is board-certified in veterinary pathology. His primary interests include pathogenesis and epidemiology of infectious diseases, avian embryonic and neonatal pathology as related to captive propagation for recovery programs, population dynamics of infectious disease, and disease risk assessment for translocation and reintroduction programs.

Ethan Schreier is vice president of advanced projects at Associated Universities, Inc. while on leave from the Space Telescope Science Institute (STScI) in Baltimore. He is a tenured astronomer at STScI and has a PhD in physics from the Massachusetts Institute of Technology. He has worked on numerous space-astronomy projects since the 1970s. As a senior staff member of STScI who helped organize the institute in 1981, he had overall responsibility for all institute activities in operations, observation support, computing, data management, and archiving for its first decade. He has filled the positions of chief data and operations scientist, associate director for operations, associate director for the next-generation space telescope, and head of strategic planning and development. Before joining STScI, he had been a senior scientist at the Harvard-Smithsonian Center for Astrophysics from its inception in 1973. His research has included the study of x-ray emission from neutron stars and black holes in binary systems and of jets and massive black holes in active galaxies. He is studying the relation of active galactic nuclei to their host environments and is a member of two large research consortia ("Great Observatories Origins Deep Survey") gathering multiwavelength survey data from most major observatories.

Patricia C. Wright received her PhD in anthropology from the City University of New York in 1985. She is a professor in the Department of Anthropology at the State University of New York at Stony Brook. Dr. Wright has served as the executive director of the Institute for the Conservation of Tropical Environments since 1992 and as the international coordinator for the Ranomafana National Park Project in Madagascar since 1987. In 1989, she was chosen as a MacArthur Fellow by the John D. and Catherine T. MacArthur Foundation; and she was awarded the Chevalier d'Ordre National (National Medal of Honor of Madagascar) by the president of Madagascar in 1995. Dr. Wright's research takes her to South America, Asia, and Madagascar, where she studies the behavior and ecology of nonhuman primates. Specifically, she is interested in monogamy, paternal care, and conservation needs, and her research has focused on owl monkeys, titi monkeys, tarsiers, and lemurs. Most recently, she has been continuing a long-term behavioral and demographic study of the Milne-Edwards sifaka (*Propithecus diadema edwardsi*), which now spans 14 years of continuous research.

Appendix B

National Academy of Public Administration: Charge to the Panel on Smithsonian Research and Panel Membership

Purpose and Scope

The National Academy of Sciences (NAS) and the National Academy of Public Administration (NAPA) have been asked by the Smithsonian Institution, OMB, and OSTP to review the Smithsonian's scientific research programs. The study objective is to examine the extent to which the Smithsonian should have a portion of its scientific research funding awarded through a competitive process.

NAS will concentrate on identifying whether research programs exist within the institution whose funding would be more appropriately awarded through a competitive grants program open to all researchers in the public and private sectors. With regard to more unique science programs not so treated, NAS will assess how best to evaluate the quality of the work. NAPA will examine how other research institutions divide research programs between in-house and competitive programs, assess the dollar implications of the NAS findings, develop alternative strategies for implementing any NAS recommendations on competition, and analyze the implications for overhead rates resulting from different strategies.

Panel Members

James E. Colvard, Chair Visiting Professor, Virginia Polytechnic Institute and State University. Former Associate Director, Johns Hopkins University Applied Physics Laboratory; Deputy Director, US Office of Personnel Management; Director of Civilian Personnel Policy, US Navy; Deputy Chief of Naval Material; Technical Director, Naval Surface Weapons Center.

C. William Fischer Former Senior Vice President for Business and Finance, Northwestern University; Executive Vice President, Brandeis University; Vice President for Budget and Finance, University of Colorado; Assistant Secretary for Planning and Budget, US Department of Education; Deputy Administrator, Energy Information Administration, US Department of Energy; Deputy Associate Director for Human Resources, and Deputy Assistant Director for Legislative Reference, US Office of Management and Budget.

Adam Herbert, Jr. Executive Director, Florida Center for Public Policy and Leadership, and former President, University of North Florida. Former Chancellor, State University System of Florida; Dean, School of Public Affairs and Services, Florida International University; Director, Northern Virginia Programs, Virginia Polytechnic Institute; Special Assistant to the Under Secretary, US Department of Housing and Urban Development.

Delores Parron Senior Advisor, Office of the Director, National Institutes of Health. Former Deputy Assistant Secretary for Planning and Evaluation at the US Department of Health and Human Services. Prior to this Dr. Parron served as Associate Director for Special Populations at the National Institute of Mental Health and Associate Director, Division of Mental Health and Behavioral Medicine, at the Institute of Medicine.

Maxine Singer President, Carnegie Institution of Washington, member of the Board of Governors and Scientific Advisory Council, Weizmann Institute of Science. Former chair of the editorial board of *Proceedings of the National Academy of Sciences*, currently member of The Human Genome Organization, and member, Board of Directors, Johnson & Johnson, former trustee, Yale (University) Corporation, and director, Whitehead Institute.

Jerry R. Schubel Visiting professor of biology and environmental studies and director of the Alternative Futures Forum at Washington College in Chestertown, Maryland. Former President and CEO of the New England Aquarium in Boston, various positions at the State University of New York at Stony Brook, including Dean and Director of Stony Brook's Marine Sciences Research Center; the University's Provost; and acting Vice Provost for Research and Graduate Studies. Also served as an adjunct professor, research scientist and Associate Director of The Johns Hopkins University's Chesapeake Bay Institute.

Project Director
Gerald Barkdoll

(from NAPA web site www.napawash.org)

Appendix C

Smithsonian Astrophysical Observatory Facilities

John G. Wolbach Library, SAO Headquaters, Cambridge, Massachusetts
The Wolbach Library staff is dedicated to providing timely information and research services to the staff at the Harvard-Smithsonian Center for Astrophysics, the Harvard University and Smithsonian Institution communities, and the international astronomical community. The library is the product of the merging of the collections of the Harvard College Observatory Library and the SAO Library and is one of the world's pre-eminent astronomical collections.

Fred Lawrence Whipple Observatory, Amado, Arizona
At the base of Mt. Hopkins in the Santa Rita Mountains, 56 km (35 miles) south of Tucson and just within the boundary of the Coronado National Forest, the Fred Lawrence Whipple Observatory is the largest field station of SAO.

MMT Observatory, Amado, Arizona
The MMT Observatory is a joint venture of the Smithsonian Institution and the University of Arizona, on the grounds of the Fred Lawrence Whipple Observatory. This innovative facility has recently been converted to house a single 6.5-m mirror.

Submillimeter Array, Mauna Kea, Hawaii
Currently under construction, the Submillimeter Array, a collaboration between SAO and the Institute of Astronomy and Astrophysics of the

Academia Sinica of Taiwan, will observe the universe at submillimeter wavelengths.

Oak Ridge Observatory, Harvard, Massachusetts
The Oak Ridge Observatory in Harvard, Massachusetts, is operated by the SAO and is part of the Harvard-Smithsonian Center for Astrophysics in Cambridge, Massachusetts.

Magellan 6.5-m telescopes, Las Campanas, Chile
The Las Campanas Observatory on Cerro Las Campanas in Chile operates twin 6.5-m optical telescopes for a consortium of institutions that include Harvard University, the Carnegie Observatories, Massachusetts Institute of Technology, the University of Michigan, and the University of Arizona.

Submillimeter Telescope and Remote Observatory, Antarctica
SAO maintains a presence on the remote continent with the Antarctic Submillimeter Telescope and Remote Observatory and plans to expand these scientific endeavors.

Appendix D

Examples of Long-Term Projects and Datasets at the Smithsonian Tropical Research Institute

RESEARCH AND MONITORING

Marine Biology

- Monitoring population of sea urchins – 15 years
- Quantifying larval settlement by coral-reef fishes at monthly intervals – 20 years

Terrestrial Plant Biology

- Quantifying flower and seed production of 625 species of trees, shrubs, and lianas at weekly intervals – 16 years
- Monitoring 62,000 trees near Manaus, Brazil, to study the dynamics of forest fragments – 23 years
- Conducting fully factorial, experimental augmentation of soil fertility (nitrogen, phosphorus, and potassium) - since 1997
- Conducting integrated series of experimental manipulations of atmospheric CO_2 concentration - since 1990

Terrestrial Invertebrate Biology

- Monitoring population fluctuations of euglossine bee population – 23 years
- Monitoring population fluctuations of over 700 species of true bugs, native bees, and the invasive African bee – 18 years
- Studying life-histories of over 1170 insect species – 26 years

Terrestrial Vertebrate Biology

- Monitoring population dynamics of 133 avian species – 23 years
- Monitoring lizard population dynamics – 31 years
- Monitoring population fluctuations of mammal species with transect counts – 21 years
- Monitoring 3000 rice-paddy fields in Senegal – 38 years

COLLECTIONS AND RECORDS

- Over 50,000 marine fossils
- Over 300,000 recent mollusks from both coasts of Central America, with precise geographic, sediment, taxonomic, and biologic information
- Extensive data on carbon isotope ratios in tropical forest plants, particularly bromeliads and other epiphytes
- Record of rainfall and temperature on Barro Colorado Island – 80 years
- Record from a class A weather station, soil moisture content, and stream flow from Barro Colorado Island – 31 years
- Record from a class A weather station, salinity, and sea level from Galeta Point
- Record of water quality in Bay of Panama – 20 years

Appendix E

Examples of Long-Term Projects and Records at the Smithsonian Environmental Research Center

RESEARCH AND MONITORING

Estuarine Biology

- Monitoring physical and chemical variables of Rhode River, a subestuary of the Chesapeake Bay – 32 years
- Monitoring water quality and tide level of Chesapeake Bay main stem – 10 years
- Monitoring phytoplankton and microzooplankton population dynamics in Rhode River – 27 years and 15 years, respectively
- Monitoring fish population dynamics in Muddy Creek with fish weir – 20 years
- Monitoring nearshore fish population dynamics in Rhode River with seining – 22 years
- Monitoring epibenthic fish and decapod crustaceans with trawls – 22 years
- Monitoring benthic infauna in Rhode River – 23 years
- Monitoring nearshore shrimp and fish with sweep-net and log-drop sampling – 11 years
- Monitoring crab, killifish, and grass shrimp survival with tethering – 11 years
- Monitoring watershed discharges and stream chemistry in Rhode River weir system – 32 years

Plant Ecology

- Studying effects of CO_2 enrichment on marsh communities and ambient CO_2 in Rhode River marsh ("World's longest-running field experiment on the effects of CO_2 enrichment on natural plant communities") – 15 years

- Studying effects of CO_2 enrichment on scrub oak at Kennedy Space Center – 6 years
- Estimating productivity in Rhode River marsh – 20 years
- Estimating productivity of old forest and mid succession forest – 25 years
- Studying deciduous-forest tree dynamics in Rhode River watershed: species composition and demography on permanent plots – 12 years

Animal Ecology

- Breeding-bird surveys in Rhode River watershed – 20 years

RECORDS AND DATABASES

- Records of precipitation, temperature, wind speed, and so on in Rhode River watershed – 32 years
- Records of precipitation chemistry (wet and dryfall) in Rhode River watershed – 26 years
- Records of micrometeorology of Smithsonian Environmental Research Center deciduous forest – 8 years
- Records of ultraviolet radiation in Maryland and Hawaii – 25 years and 16 years, respectively
- Geographic database of topography, land-use composition and patterns, streams and rivers, and shorelines of Rhode River watershed – 55 years
- Geographic database of land-use history of southern Anne Arundel County based on historical surveys and records, aerial photography, and satellite imagery – 300 years

Appendix F

List of Acronyms

ARI	Interoceanic Canal Authority
AURA	Associated Universities for Research in Astronomy
AZA	American Zoo and Aquarium Association
CAL	Conservation Analytical Laboratory
CBCES	Chesapeake Bay Center for Environmental Studies
CCI	Canadian Conservation Institute
CRL	Conservation Research Laboratory
CXO	Chandra X-Ray Observatory
DARPA	Defense Advanced Research Projects Agency
DOC	Department of Commerce
DOD	Department of Defense
DOE	Department of Energy
DOI	Department of the Interior
EPA	Environmental Protection Agency
ELIPSE	Environmental Latino Initiative Promoting Science Education
GCI	Getty Conservation Institute
GIS	Geographical Information System
INAA	Instrumental Neutron Activation Analysis
LTER	Long Term Ecological Research
LTREB	Long Term Research in Environmental Biology
MBL	Marine Biological Laboratory
MMT	Multiple Mirror Telescope
NAIC	National Atmospheric and Ionospheric Center
NAPA	National Academy of Public Administration

NAS	National Academy of Sciences
NASA	National Aeronautics and Space Administration
NCAR	National Center for Atmospheric Research
NIH	National Institutes of Health
NMNH	National Museum of Natural History
NOAA	National Oceanic and Atmospheric Administration
NOAO	National Optical Astronomy Observatories
NRAO	National Radio Astronomy Observatory
NSF	National Science Foundation
NZP	National Zoological Park
OMB	Office of Management and Budget
OSTP	Office of Science and Technology Policy
PAEC	Performance Accomplishment Evaluation Committee
SAO	Smithsonian Astrophysical Observatory
SCMRE	Smithsonian Center for Materials Research and Education
SERC	Smithsonian Environmental Research Center
SI	Smithsonian Institution
SI/MAB	Smithsonian Institution/Man and the Biosphere Biological Diversity Program
SMA	Submillimeter Array
STRI	Smithsonian Tropical Research Institute
STScI	Space Telescope Science Institute
SWAS	Submillimeter Wave Astronomy Satellite
UCAR	University Corporation for Atmospheric Research
USAID	United States Agency for International Development
USDA	United States Department of Agriculture
USFWS	United States Fish and Wildlife Service
USGS	United States Geological Survey
VLBI	Very- Long- Baseline Interferometry
WHRC	Woods Hole Research Center